THE NORTH CAROLINA FLOOD

JULY 14, 15, 16, 1916

W. M. BELL

Published by Left of Brain Books

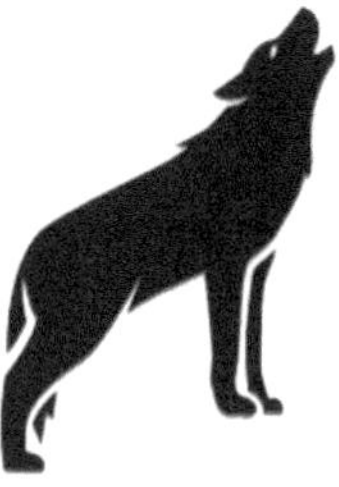

Copyright © 2021 Left of Brain Books

ISBN 978-1-396-32060-6

First Edition

Table of Contents

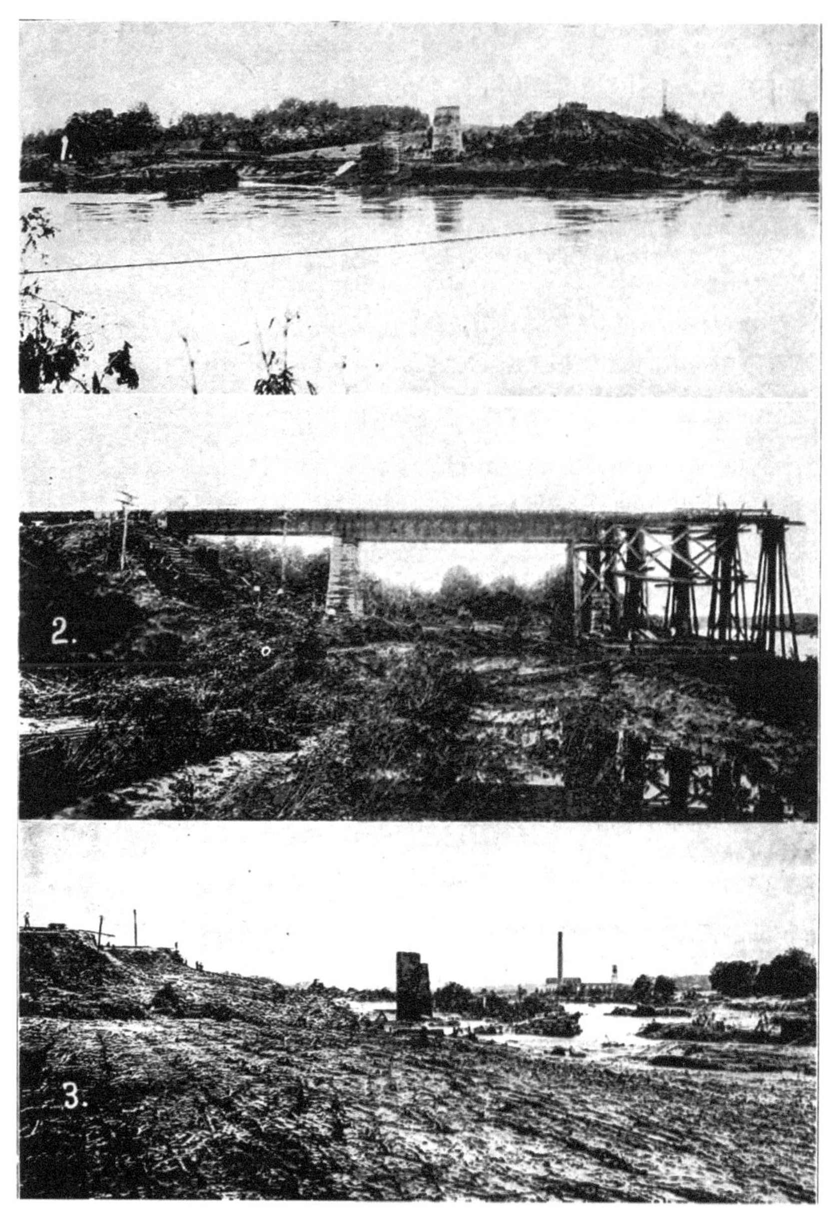

BRIDGES WASHED AWAY ON THE CATAWBA, NEAR CHARLOTTE.
Abutments left of Southern railway bridge on Gaston county side. (2) Mecklenburg side of same bridge. (3) What remains of Seaboard and P. & N. bridges at Mount Holly. Showing fields of corn and all trees and banks swept away. (Photo by Cushman, Charlotte.)

THE STORM WHICH CAUSED THE FLOOD

By O. O. ATTO
U. S. Weather Bureau, Charlotte

The variety of disturbance which caused the excessively heavy rains in western North Carolina July 14–16, 1916 is called by meteorologists, "Tropical Hurricane" or more often, "West India Hurricane". This is very destructive kind of whirling storm resembling much the well-developed low pressure area over the land, but being much more pronounced in its characteristics.

Storms of this type usually form to the south and southwest of a sort of semi-permanent area of high pressure which occupies the middle Atlantic Ocean during the summer and early fall months. As the season advances, they develop farther and farther to the westward along the southern limit of the northeast trade winds, and usually between 8 and 20 degrees north latitude.

The near approach of a hurricane is indicated by the appearance of thin, upper clouds, and a long rolling swell of the ocean. After a slight rise, the pressure begins to decrease steadily, the clouds thicken, then the pressure decreases more rapidly, and squally wind and rain begin, both increasing in intensity as the centre of the storm approaches a given place. With the pressure decreasing rapidly, high winds and heavy rains continue, often for many hours. Then the sky suddenly clears, and the sun comes out, and the unsuspecting think that the storm is wholly past. But not so. The sky soon becomes darkened with dense clouds, the wind shifts and rises to hurricane velocity, and torrential rains again begin. These conditions prevail for several hours more, but with rapidly rising barometer. At the close of the storm the cirrus or upper clouds are again visible, and nothing skyward indicates that a severe storm has passed. This kind of storm ranges in size from 500 to 1000 miles in diameter, and the storm progresses from one place to another at about 15 to 18 miles per hour.

On passing from the ocean to an extensive land area, the hurricane usually retains its destructive characteristics until the storm is wholly inland, where it generally decreases in strength and often entirely disintegrates. When the center of one of these storms moves inland and northeastward over the Atlantic coast states, places on the coast will experience destructive wind velocities, extremely high tides, and excessively heavy rainfall. Should the storm move to the Atlantic coast but with its center over the ocean, places on the coast will experiment westerly winds, and the tides will not be so high nor the rainfall so heavy, as in the former condition.

The first wireless report of the storm of July, 1916, as received by the Central office of the Weather Bureau at Washington, indicated its center to be nearly 400 miles almost due southeast of Jacksonville, Florida, on the morning of the 12th. The following morning it had moved about 200 miles in the direction of Charleston, S.C. and at 8 a. m. on the 14th, its center moved inland almost directly over the city of Charleston. Thence it moved slowly northwestward till it approached the mountains, where it re-curved to the northeastward, passing over the western portion of North Carolina and Virginia, decreasing all the while in intensity.

This particular direction of motion, produced ideal conditions for the heavy rains which it caused. Such a motion resulted in northeasterly winds which shifted to the east with the near approach of the storm and to the southeast after its center had passed. These winds, coming from off the Ocean, were consequently laden with vapor, which condensed on approaching the elevation of the interior, with the resultant torrential downpour.

The rainfall at Charlotte during the passage of the storm totaled 5.15 inches, with a maximum 24-hours fall of 5.04 inches, this being a new record for 24-hours rainfall in a period of 38 years.

STORM AND FLOOD REVIEWED FROM CHARLOTTE

By W. M. BELL

More than twenty-five years ago Bill Smith, an old man, lived in Yadkin county, near Elkin. When he worked he was a well-digger, but most of his time was spent around town. He delighted in telling of great things that happened in the country "before the war" and, during the war, especially of how the country was destroyed by Stoneman and his army on its march through that section. The old people living there said lots of Bill's yarns were true, only "he did not tell it bad enough." The children, who listened to his stories, could hardly believe all he said and named him "Lying Bill" Smith.

That was almost thirty years ago.

Thirty years from today the children, not yet born, will listen to the old men—one of them may be named Bill Smith—telling about the great hurricane and flood that swept the mountain country of North Carolina July 14, 15, and 16, 1916, and the description of this great flood will sound so unreasonable that they, too, will be called "Lying Bill" Smith.

IT CANNOT BE DESCRIBED.

It happened suddenly. The wind blew; the clouds divided and poured torrents; earth and boulders that had been resting on the mountain sides for hundreds of years went tearing down, carrying death and destruction in their wake; branches, creeks and rivers became roaring torrents sweeping down the valleys taking everything before them. It was all over in three days but in that time a damage of millions of dollars had been done to farms, railroads and industrial plants and a hundred and more men, women and children had been sent to eternity, some of them swept away and never being heard from.

Railroads and factories were rebuilt but many of the farms, ruined and washed, will remain as marks of the great flood for time to come.

THE BEGINNING OF THE STORM.

Early in the week of July 10th, the Government Weather Bureau at Washington, issued warnings stating a storm was forming along the South Atlantic coast. It was expected that Charleston, S. C., would be the center of the catastrophe. It did strike Charleston but only small damage was done to that city and section. Instead of keeping to the coast the storm turned inland and Piedmont and Western North Carolina was in its direct path.

AT CHARLOTTE.

Thursday morning, July 13th, rain began falling in Charlotte and a brisk wind was blowing. It continued to rain throughout the day and night. Friday it was raining hard, the wind was rising, and by nightfall a deluge of water was falling. The wind was ripping around the corners of buildings and in the open, at a velocity of more than 50 miles an hour. During the night awnings were torn from buildings, signs ripped from their holdings, plate glass store fronts smashed, roofs blown away, homes and stores flooded with water, and beautiful trees uprooted and those that remained standing shorn of their limbs in many instances and strewn along the streets for blocks. Next morning the city looked like a "cyclone had struck it." This was Saturday morning, July 15.

But Charlotte and vicinity had been let down easy. Soon reports began to be received that the mountain sections were in the grip of such a storm as had never before been experienced. Enquiry at the weather bureau brought the information that the storm was central over Asheville. Telegraph and telephone connections with all sections west of Hickory were down and only a guess of what was going on farther in the mountains was left for those who had been interesting themselves with the path of the storm. A dispatch from Hickory at 7 p. m., stated that the rainfall there had been more than ten inches in the ten hours and that rain was still falling in torrents. Soon afterwards the wires went down.

Saturday night at midnight when the Southern trains arrived, running late, under "caution" orders, trainmen reported the Catawba river rising several feet an hour. One engineer stated that when he crossed Broad river, near Spartanburg, it was a regular ocean, and that when he reached the Catawba it was a raging torrent. He shook his head when enquiry was made as to the rainfall up the river. He was an old trainman and, though he did not say so, he had in mind no doubt, many troubles that had already befallen the people farther up the river.

Later developments proved that this old trainman's thoughts were right. At that very moment a devastating flood was tearing down the mountains sweeping everything before it and men were risking their lives in the rescue of any unfortunates caught in the quick rise of the waters. Sunday morning the Catawba was running well over twenty feet high and still rising.

At the Seaboard Air Line and the Piedmont & Northern railway bridges, near Mount Holly, thousands of people had gathered to watch the flow of the river and its rapid rise. That thousands of dollars worth of property had been destroyed up the river was evident from the quantities of all kinds of drift passing. Heavy timber, parts of bridges, parts of houses, bales of cotton, logs, big trees torn from their roots, hay stacks, wheat shocks and various other kinds of drifts floated by. Live stock, chickens and rabbits were seen to go by on all kinds of rafts.

In the afternoon the big iron bridge of the Seaboard was forced out and down with a crash, hundreds of people standing on its approaches fleeing for safety. Later the new, modern, steel bridge of the Piedmont & Northern, standing five feet higher than the Seaboard, was forced to follow and went down, slipping easily from its piers and with it six heavy loaded cars of coal that had been run out on it for the purpose of holding it down. It was dark when this bridge went down and the heavy power lines and trolley wire carried with it, heavily charged with electric current, lighted the heavens as they came in contact with the water. It was a show not soon to be forgotten by the onlookers. Three handsome highway bridges, one a beautiful structure recently completed at a cost of $100,000 to the county of Mecklenburg, had earlier in the day been swept away.

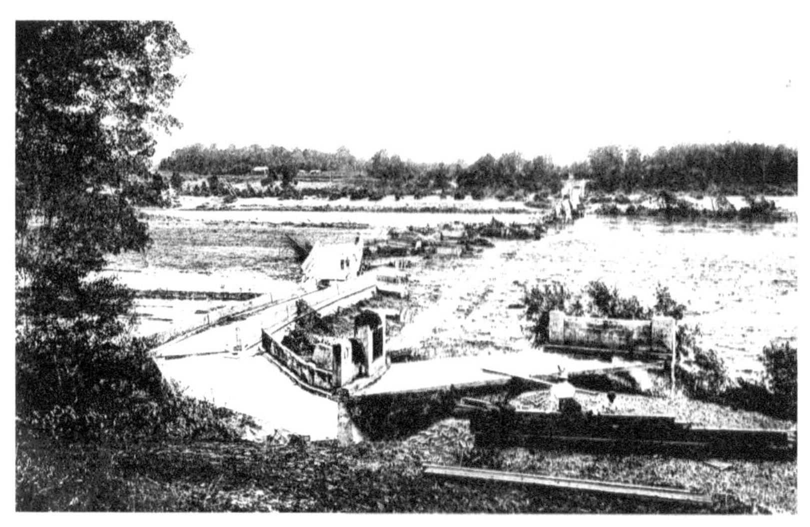

Wreck of Highway Bridge Connecting Mecklenburg and Gaston Counties.
At Belmont. Was Built Two Years Ago at a Cost of $100,000.
(Photo by Cushman, Charlotte.)

At the Southern Railway Bridge.

The heavy iron bridge of the Southern Railway at Mayesworth, three miles above Mount Holly, at 5:30 Sunday afternoon was the scene of the greatest catastrophe recorded. When this bridge was swept away it carried down eighteen or more employees of the railroad who had been doing heroic work for hours trying to save the structure. The crash came without warning and the men were thrown into the rushing waters, six of them never to be seen again. That some of the men had managed to catch to trees farther down was known from the cries for help that could be heard above the roar of the water. It was then dark and to make any effort at rescue would have been simple suicide. Monday morning the river had risen more than fifteen feet and the trees that had held the unfortunate men were either submerged or carried away.

Men Who Went Down With the Bridge.

The men known to have gone down with the Southern Railway bridge included Resident Engineer Joseph Killian, H. P. Griffin, C. S. Barbee, R. O.

Thompson, W. L. Fortune, G. C. Kale, J. N. Gordon, C. W. Kluttz, and H. O. Gulley, all white, and ten or more negroes, all employees of the Southern. The white men with the exception of Gordon, Barbee and Kluttz were rescued the following day and a number of the negroes. Gordon's body was taken from the river two days later and Barbee's was found on Friday. The body of Kluttz was never found.

Of the heroic rescues of the men marooned in trees along the river graphic stories might be written. Early Monday morning B. M. English and H. T. Vernor put out with a boat for Thompson who was hanging to a tree near the middle of the rushing stream. They reached him and had succeeded in getting him in the boat when, in his delirium, he capsized the boat and threw all three men into the water again. Vernor saved himself and both the other men by drawing them to a tree farther down. They were marooned for more than nine hours before being rescued by two negroes.

Alphonso Ross and Peter Stowe, two negro men, reared on the river and familiar with manning a boat volunteered to make the effort at rescue. A boat was hastily made and the men went out. Trip after trip was made until they had brought to the shore Vernor, English, Thompson, Killian, Gurley and Kale.

For this act of bravery a purse of $550.00 was presented to them a few days later. The purse was collected through the Charlotte Observer and contributed to by both white and colored citizens.

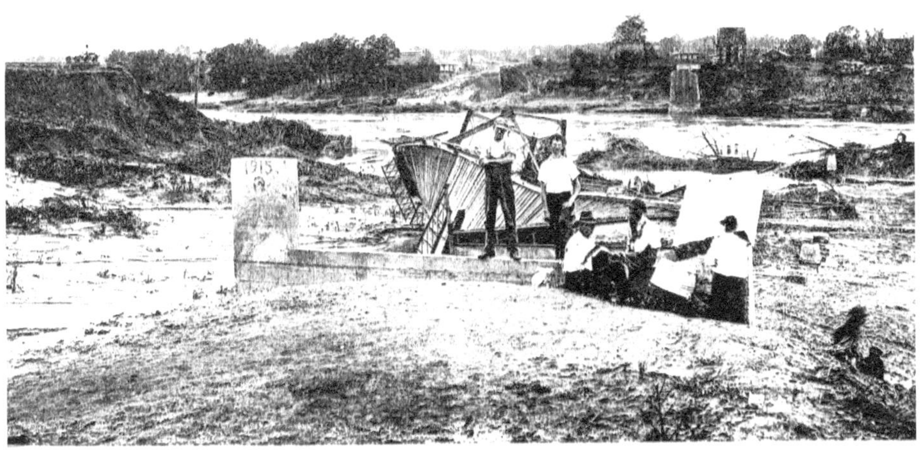

Wreck at Mount Holly. Showing Destruction to Highway, Seaboard Railway and P. & N. Railway Bridges. Property Loss $200,000 and More.
(Photo by Cushman, Charlotte.)

MEN NEARLY DELIRIOUS.

Messrs. Killian, Thompson, Gurley and Kale had been marooned for twenty hours struggling every moment against a relentless current for their lives. They were dazed and stupefied. Their throats were sore, their eyes swollen and their heads were roaring the merciless music of the Catawba's cantations. Their shoes were gone and they were shivering from exposure. They were wild with excitement. Their eyes carried the faraway look of those "who had gone down in the deep." They were a vision that was painful to behold. When they were brought to the bank tears filled many an eye and the speech of the men was interrupted by the swelling of emotions within their own breasts. The men were brought to Charlotte, given medical attention, and sent to their homes and families.

ON DOWN THE RIVER.

Farther down the Catawba continued its destruction. At Fort Mill, S. C., the 510 foot steel bridge of the Southern was swept away cutting off rail connections over both main lines leading to the South from Charlotte. The Mountain Island cotton mills at Mount Holly was swept away and with it a number of homes of operatives and 800 bales of cotton. Nothing could stand before the onrushing waters. The river was at that time—it had reached its crest—51 feet, some said, above its ordinary level.

AT THE CATAWBA'S HEAD.

The Catawba rises in the mountains to the right of Asheville. It flows down one of the most picturesque valleys of the western part of North Carolina; it is fed by numbers of smaller rivers and creeks. All along its banks, in McDowell, Burke, Catawba, Caldwell, Iredell, Lincoln, Mecklenburg and Gaston counties, the finest farms of the state are to be seen and they are in the highest state of cultivation. Many of these farms, especially in McDowell, Burke, Caldwell and Catawba counties have been reduced to sand and gravel. Dotted along the river are great hydro-electric power plants and great concrete dams. None of the power plants were washed away but all were buried in water and sand and damaged thousands of dollars. Breaks occurred in a number of the dams that will cost thousands of dollars to repair. Located along the river were beautiful, modern cotton mills, the pride of the section and representing investments of thousands. A majority of the mills stood the flood pressure but were overflowed and greatly damaged. One mill near Statesville was completely wrecked and another one on the opposite bank, owned by the same corporation, was damaged 30 to 50 percent. All of them had warehouses of cotton washed away, representing large amounts of money. All these losses were complete as no insurance covered such losses.

DAMAGE TO RAILROADS IN THE MOUNTAINS.

The Southern Railway from Statesville west towards Asheville was the heaviest loser when railroads are considered. An article appearing in this book by Mr. F. C. Abbott, of Charlotte, gives some idea of the damage to this one line and the loss to the corporation. Not a railroad bridge was left standing where the road crossed the river and in many places slides and washouts occurred miles away from any stream, so great was the rainfall in that section. Trains caught between bridges and slides were forced to stand for days before lines were reconstructed. Passengers marooned on these trains told vivid stories of their experiences cut off from the world and not knowing the conditions on down the mountain. In some instances only limited food was to be had and no lights or water. One passenger stated that he never got so much in his life for a dollar-and-a-half. He was in a Southern Railway train from Saturday morning until the following Wednesday, living in a Pullman that was lighted at night by a tallow candle.

The Carolina and North Western Railway was also a heavy looser. This road reaches from Chester, S. C., to Lenoir, with an extension of 30 miles from Lenoir to Edgemont for lumbering. The greatest damage to this line of road was from Newton to Lenoir and to Edgemont. Bridges were swept away and track and roadbed damaged so that for days traffic was suspended. Beyond Lenoir towards Edgemont almost the entire road was carried away. This section was in the center of the cloudburst and rushing waters carried railroads, saw mills, farms, farm houses, and everything before them. The sweep of the water in this section at one place is said to have completely washed away a cemetery that had been in use for more than fifty years, leaving not a mark of a grave. Many lives were reported lost and untold damage was done to property.

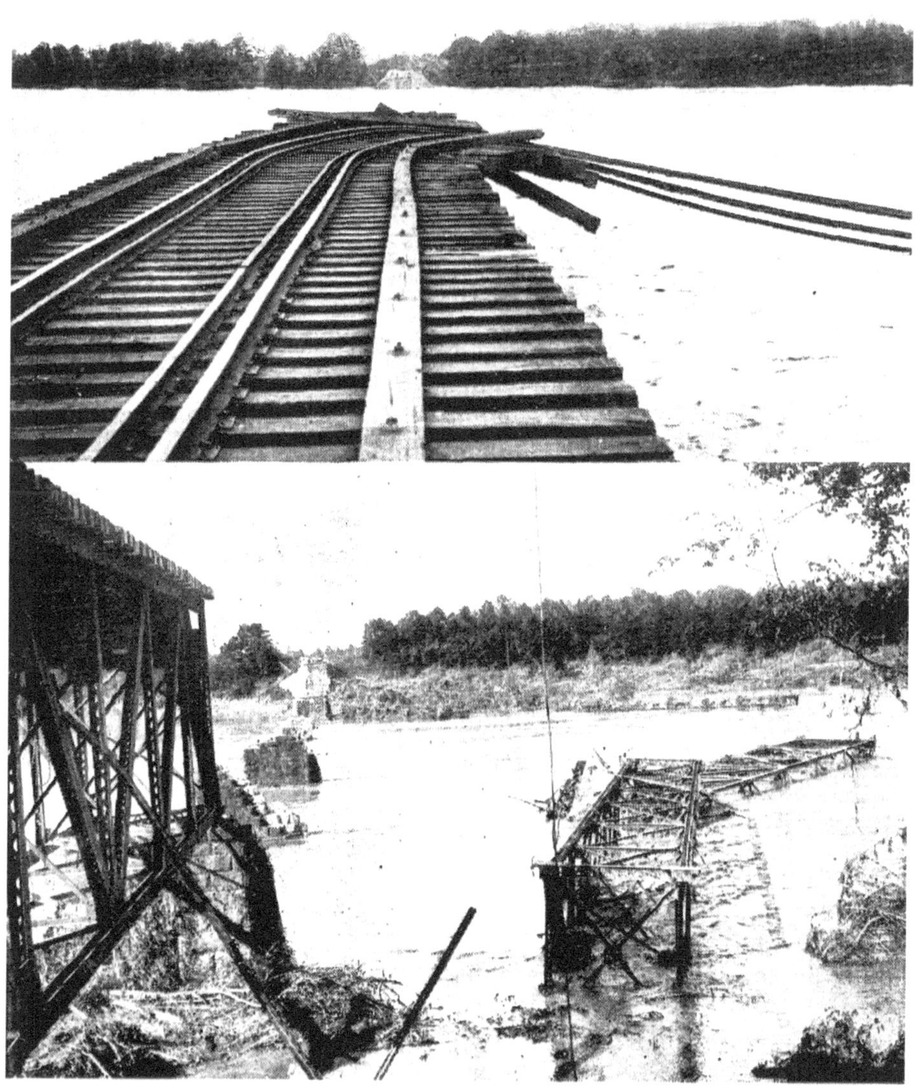

SOUTHERN RAILWAY BRIDGE, CATAWBA RIVER, NEAR ROCK HILL.
This bridge (510 feet) was the last Bridge to go out breaking the main line of the Southern from Charlotte to Jacksonville. Upper view shows river at flood tide; lower, wreckage after river receded. (Cut courtesy Rock Hill Magazine.)

In Alexander county, near Taylorsville, certain sections came in for great property damage. Flood waters of the streams caused the washing away of the cotton mills at Liledown and Alspaugh and every bridge and corn mill along the streams. Away from the streams cliffs and boulders broke away from their holdings and went crashing down the mountains carrying death and destruction. Over in Wilkes county and farther up in Watauga and Ashe counties similar conditions obtain.

SEABOARD AIR LINE LOSES HEAVILY.

The Seaboard Air Line railway from Charlotte west to Rutherfordton was loser principally from bridges being swept away. Its heavy iron bridge, near the P. & N. bridge at Mount Holly was one of the first to go out on the Catawba near Charlotte. Damage from washed roadbed and culverts was reported all along and for several weeks after wards traffic was delayed. Its line east from Charlotte to Monroe was never closed but farther south towards Columbia and Savannah traffic was temporarily suspended on account of the high water and dangers from moving trains over weakened bridges.

The Southern Railway was also heavily damaged on its lines from Asheville to Spartanburg, down the Saluda mountains and its lines along the Yadkin river from Wilkesboro to Donnaha a distance of fifty miles or more. Accounts of the great damage and the trail of the flood in these sections are told of in special articles from eyewitnesses, published in other sections of this book.

C. C. & O. RAILWAY ALSO HEAVY LOSER.

The Carolina, Clinchfield & Ohio Railway stretching from Dante, Va., to Spartanburg, S. C., was also a heavy loser in the mountains. This road starts across the mountains at Marion, N. C., and winds around cliffs and mountains and through tunnels for miles until the Tennessee line is reached. Its greatest damage was due to mountain slides and cloudbursts of water sweeping down carrying rails, ties and roadbed with it to the chasms below. The monetary damage to this road could not be estimated. It was closed for several weeks afterwards and it will be months and perhaps a year before the road is back in as fine shape as before the flood. It was the finest built road in the North Carolina mountains, with heavy rails and rock ballast.

On the Western Slope of the Mountains.

The great rain fall being central over the Asheville section it carried the flow of the water west as well as east. New River, Tennessee, Toe and other rivers and streams on the western slope while not as high as those on the eastern side did great damage and overflowed. All along the valleys thousands of dollars in damage was done to farms, bridges and industrial plants. At Knoxville the river ran well up into certain business and residential sections causing no loss of life but great property damage.

New river that which borders Northwestern North Carolina and southwest Virginia was a raging torrent. It made a record for flood stage, sweeping what is known in the mountains of North Carolina as "New River Valley" clean of everything in its path. At Fries, Va., the half million dollar cotton mills—The Washington Mills—was greatly damaged and the great concrete dam swept away. This property is largely owned by North Carolina capital. At East Radford, Va., thousands of dollars was lost by damage to property along the river. The Norfolk & Western Railway was the heaviest individual loser along this river. This railroad corporation has been building short lines of road back into the mountains of North Carolina and Virginia for several years and these roads were the lines most damaged in the New River Valley.

The Loss the Section Over.

No estimate within a million dollars can be reached as to the loss to property sustained from this great flood, not considering the loss of life and the loss from interrupted business. Railroads and bridges are already largely rebuilt but the farms will remain barren in many sections forever. No soil is left, only barren rocks. This is true especially in the mountain sections. The natives ordinarily of scant means, many of them barely able to eke out an existence will be forced to emigrate to new fields. It is true the Governor was prompt to issue an appeal for these stricken people and the citizens quick to respond but the funds, no matter how great, will never be able to restore to them the lost lands on which they made a living.

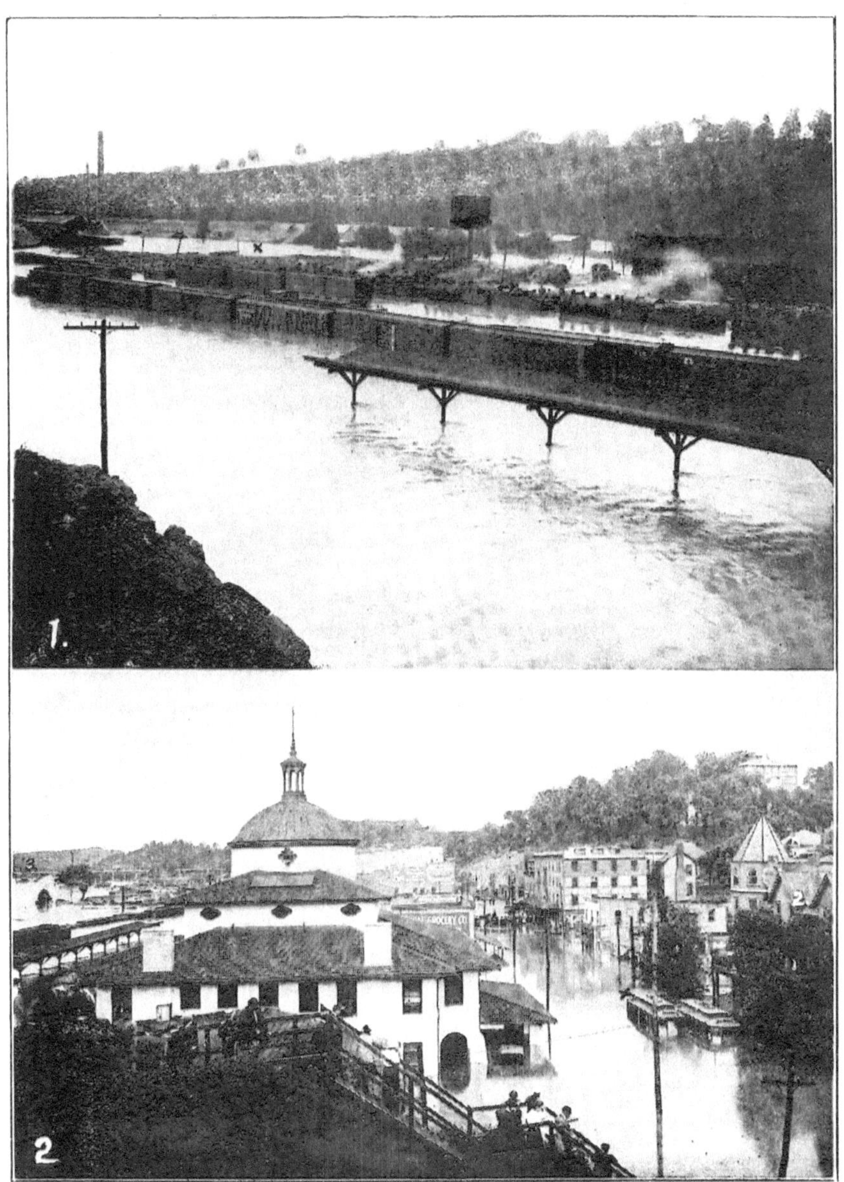

SCENE AT ASHEVILLE.

(1) Southern railway yards, X locates Hans-Reece Tannery under water. Great property loss here. (2) Southern railway passenger depot under water. X shows where two men were drowned trying to get food to marooned guests in Glen Rock Hotel.

ASHEVILLE, BILTMORE AND NEARBY POINTS

By HELEN C. BLANKENSHIP

FIRST RUMORS OF THE STORM.

On Wednesday, July 12, 1916, the following little news item was sent out by the Associated Press and appeared in the daily papers of the country:

WEATHER DISTURBANCE.

"Washington, July 12—Weather bureau reports today give some indications of a disturbance in the extreme eastern Caribbean Sea."

The item appeared in the Asheville papers in an inconspicuous position, read by few, remarked, probably, by none. Could it have been given its true news value in relation to the proportions it was to assume within the next few days, it would have been printed in scare head type over the biggest part of the front page, would have been on every lip. For this was the cloud, no bigger than a man's hand, which was to deluge Western North Carolina with such horror, ruin and grief as the peaceful and beautiful region had never in all its history known before.

Thursday the "follow-up" item said, "The Caribbean sea disturbance was apparently central this morning near the south of Porto Rica, according to weather bureau reports. Its intensity was still unknown."

Friday afternoon the coast of South Carolina and Georgia was swept by a hurricane, government storm warnings were ordered up, and the hurricane was felt in central South Carolina, the wind increasing in violence.

At Asheville, the French Broad had been high all week. All week it had been raining. Curious sight-seers had thronged the concrete bridge to see the river lapping the front steps of little houses ordinarily several hundred feet from the water's edge. But nothing had been thought of it. The Asheville rivers had never in the history of the oldest inhabitant done any damage, caused any apprehension. When Saturday afternoon the Asheville evening paper carried the story that the Swannanoa was at flood tide, and that great timbers and piles of lumber were sweeping down the stream and threatening the house of J. C. Lipe, near the river, the account was read with passing interest only. "It is reported that there has been a cloud burst on the head waters of the Swannanoa," said the paper. A cloud burst indeed! It was in this form that the rumor of what was happening in the "Big Mountains" first reached Asheville.

The tropical hurricane, had traveled north and west from South Carolina. Borne along by a wind of irresistible power, the vast mass of clouds, heavily-laden with water, had crashed against the peaks of the Blue Ridge, and had poured upon their slopes deluge enormous beyond the power of man to imagine or describe.

At Alta Pass, in Mitchell county, where the center of the storm burst, and where the rainfall was heaviest, 22.22 inches of rain fell in the 24 hours preceding 2 p. m. on the fateful Sunday.

At Black Mountain, as said W. F. Randolph, of Asheville, who was at his summer home, Orchard Camp, the rain came down in sheets, in streams, in tubfuls.

At Old Fort, at the foot of the Blue Ridge, as the Old Fort Sentinel stated, rivulets became streams, streams became creeks, creeks became rivers, and graphic phrase, "rivers became frightful."

Capt. John T. Patrick stated that at Bat Cave and Chimney Rock, the noise of the rain beating on the roofs of the houses was only comparable to thunder.

All this rainfall, be it remembered, fell upon ground already soaked and saturated down to bedrock, upon streams already bank-full, due to the storm of the preceding week. No wonder then, that rivers were doubled, tripled in volume, with resulting destruction widespread and terrible.

DAWN OF THE FATEFUL SUNDAY.

Asheville was on the outskirts of the path of the storm, and Sunday night fell with the people ignorant of the calamity that was approaching. Laborers took the street car to their suburban homes, parties motored out to their country places, the trains pulled out of the depot as usual, carrying their human passengers bound for points far and near. Saturday night, rainy and windy, darkness fell, and people went to bed and to sleep—to sleep, many of them, till late hours Sunday morning.

The blowing of the whistle of the cotton mill, and the ringing of the riot call on the fire bell, awakened them to the strangest Sunday Asheville had ever known. To those who lived beyond the sound of these alarms, the first inkling of the situation came when they found they had no electric lights, and no gas, and that there were no street cars running. People began to question their neighbors, to telephone, to stop passersby, and soon the news was spreading like wildfire all over the city and suburbs, and to nearby villages. The wildest rumors were current, though most of them were no worse than the truth.

"Asheville and Biltmore are flooded!" the cry went around. "The water is up to the ceiling in the depot. It is six feet deep in Dr. Elias' house in Biltmore. It is in All Soul's church—it is in the Vanderbilt hospital—the beds are floating—the patients are drowning! The tannery is washed away—bridges are gone. Captain Lipe and some of the nurses are drowned at Biltmore. Other people are up in trees, surrounded by water, and they cannot get them out of the river. The Swannanoa is a mile wide! Box cars are floating down the French Broad. The lakes at Hendersonville have broken, and hotels from Hendersonville have floated down the French Broad. Street cars are under water in front of the depot."

As the news spread, the streets leading to Biltmore and the depot, to the concrete bridge and Riverside Park, became thronged with a steady procession of men, women and children on foot, of autos, and carriages, and of horseback riders and wagons from the country, and the work of life-saving which was going on at Biltmore and in the river district of Asheville was watched by a vast crowd.

RESCUE WORK AT BILTMORE.

At 4 o'clock Sunday morning flood torrents had burst without warning into the village of Biltmore; the rising waves drove the people to the hills for miles along the Swannanoa and French Broad, and houses were tossed in the waves like egg shells and lashed in pieces against the concrete bridges.

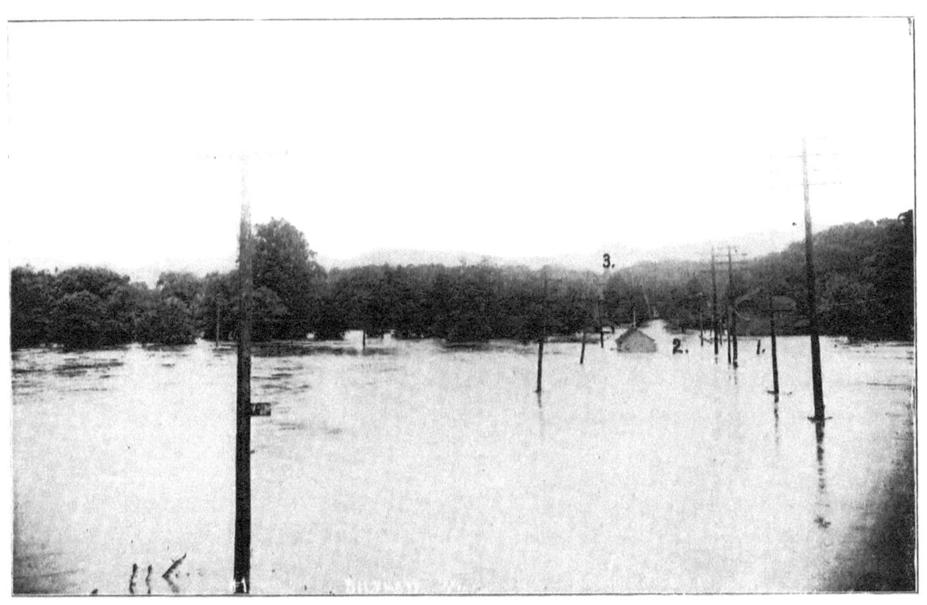

BILTMORE UNDER WATER.

George Vanderbilt's village, near Asheville. (1) J. C. Lipe's home. Mr. Lipe and three young ladies drowned trying to escape. House afterwards went out. (2) Biltmore bridge under water. (3) All Soul's Church. (Mrs. Vanderbilt's church.)

The principal interest centered from an early hour in the morning until late afternoon about the Lipe house of Biltmore where Captain Lipe, and the nurses, the Misses Walker and Miss Foister, lost their lives.

P. A. Miller, mayor of South Biltmore, was an eye-witness to the entire scene. Here is his story:

"My little boy woke me about 6 o'clock," he said, "saying that the river was up and Captain Lipe's family in danger. I went right out there. Captain Lipe was up in a tree near his house, holding his youngest daughter, Miss Katherine Lipe, above him. Miss Charlotte Walker, and Miss Foister, nurses from Biltmore hospital, and Miss Louise Walker, Miss Walker's sister, were standing at the foot of the tree in water up to their necks. They were holding to the tree and at times tried to climb up into it.

"Everyone of the Lipe family, and the nurses, had once gotten out of the house at 5 o'clock in the morning in water up to their ankles. They did not believe the water would rise any more and went back after their belongings. The water caught them so suddenly that they could not get away.

"One by one the victims gave way, let go their hold and sank immediately. A young man was swimming to the last of the young ladies with a rope when she turned loose and sank. Captain Lipe was the last to turn loose. He had been in that cold water for six or eight hours, with the river lashing his back and beating him against the tree, when he gave way and fell into the water. He was seen to go ten feet, to sink, come up, go under again and was never seen any more.

"He left his daughter, Miss Kathleen still clinging to the tree. She stayed like that some two hours when a young man swam to her and went up the tree. Another young man swam out and took her a rope. They tied her up in the tree, well above the water, so that her weight was suspended by the rope under her arms, before they got a boat to her. We had phoned to Skyland in the morning for a boat, and young Frady brought it to Biltmore on a wagon. Raymond Plemmons, Mrs. Vanderbilt's chauffeur, and Will Donnahoe who works at the Vanderbilt house as footman, got the boat to her—a flat bottomed home-made boat, and rowel to shallow water. Dr. Elias met her at the water, and we carried her a quarter of a mile through the woods to Dr. Smather's machine, which took her to the hospital. She kept asking us not to hurt her left arm, and said that she was beat to pieces against that tree. It was about 2 o'clock when she was rescued.

"All this time, Mrs. Milholland, Captain Lipe's oldest daughter was in the water near the ball park pavilion, holding to a tree. Another man was there, too, whose name I do not know. Mrs. Milholland had been out of her house and returned. She phoned Mr. Fiezer, the livery man to come get his horses out of the stable and she got caught by the water herself. My brother, A. A. Miller, took the first rope to them. He swam from the office building, was washed under, came up again, but lost the rope. Walter Curry took the second rope. He stopped at the tree above, and threw it to them. He went under, but they got the rope. He caught the tree and stayed there for two hours, then swam out at the lodge gate.

"The man lashed Mrs. Milholland to the three. In a few minutes about two wagon loads of lumber, solidly packed, came down the river right against their backs, and settled against the tree. The man untied the rope, climbed upon the lumber, rested, then pulled Mrs. Milholland up, she rested, and then he lashed her to the tree two feet above the water. They were in the tree four hours. They got them out about 3 or 4 o'clock. The water began to fall at 12.

"In the meantime some Biltmore carpenters, Mr. Hamilton, Mr. Hall, and Mr. Creasman, had made a boat. We used this to rescue families along Brook street, among them the family of Ben Taylor, also Mr. Weldon, the night operator at the Biltmore depot, who had been in water up to his neck since 3 o'clock."

But for the bravery of C. P. Ryman and R. Ball, Miss Kathleen Lipe would have undoubtedly perished in the flood. Like true heroes, these men after risking their lives, said nothing about it.

Sunday noon, while Miss Lipe was still clinging in the tree from which four had been swept to a watery grave, Ryman and Ball, unknown to anyone, were constructing a small raft. When finished, they pushed it into the water and started for the tree, but when the frail raft was within a few feet of the tree it struck a lamp post and the men were thrown into the water. Mr. Ryman succeeded in swimming to the tree to which the girl was holding, but Mr. Ball was carried on down stream and finally succeeded in getting a foothold on the lodge gate where he remained for five hours.

Mr. Ryman, on reaching the tree lifted Miss Lipe as far out of the water as possible, then tied a small rope about her waist and fastened her securely to the tree. He then climbed the tree, and reaching down released the rope about the girl and attempted to pull her up into the tree, but he was too exhausted after his battle with the swift water and was compelled again to tie the girl to the trunk of the tree. He then climbed into the tree and for five hours stood ready to plunge into the water should the girl be torn from the tree.

Ryman was taken from the tree by the men who later reached him and Miss Lipe with a boat.

Will Cooper, also of Biltmore, tied a rope about his body and swam, battling the swift current, all the way from Biltmore office to the tree where Miss Lipe was clinging, which was near the estate lodge gate.

Ryman and Ball put on in a boat they had constructed in an attempt to reach the girl. When a short distance away, their craft was hurled against a lamp-post and mased. The pair swam to the tree harboring Miss Lipe and drew themselves into the branches.

Cooper witnessed the disaster, put a rope around himself with a knot he could easily loose and reached the tree in a record breaking swim. Once there, he helped Miss Lipe further into the branches, bound her fast to the trunk and tossed the end of the line to Ryman.

Almost exhausted, he dropped back into the flood and was carried swiftly toward the lodge wall. He attempted to reach the building but the current was too much for him. Fifty feet below he caught hold of some bushes and hung fast.

The lodge keeper, Franks, seeing his plight, tied several sheets together and let the improvised life line drift down as far as possible. When Cooper got his wind, he pulled himself along by the bushes to it and was dragged to safety.

If it had not been for his ready wit and courage in the face of terrific danger, it is probable that Miss Lipe never could have hung on until a boat rescued her.

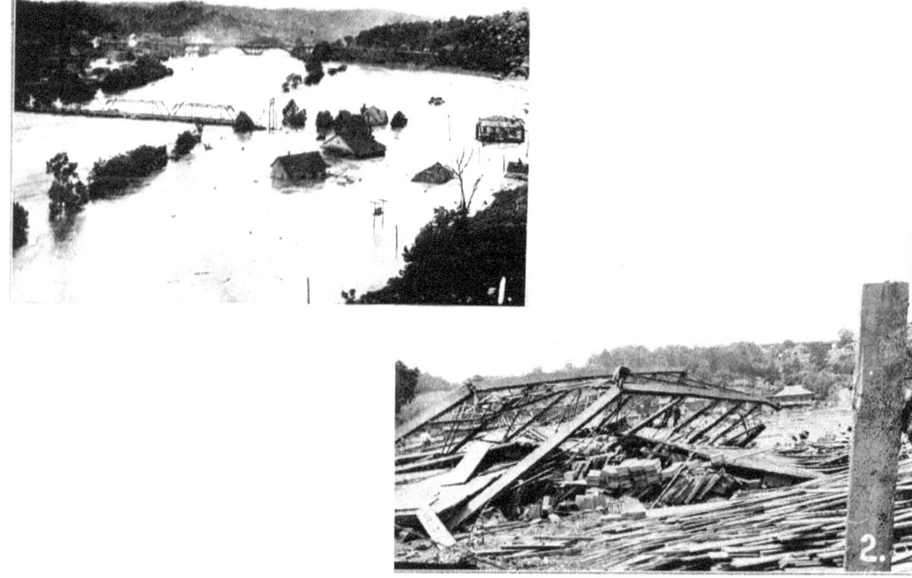

SMITH'S BRIDGE AT ASHEVILLE DURING AND AFTER THE FLOOD.

A LAME GIRL SAVED—STORY OF MISS LIPE.

There was nobody along the entire waterfront better placed to view the enormous destruction in life and property than Miss Nellie R. Lipe. The fact that she is lame and partially unable to walk about freely added to the terror of her position on the day her father was swept to his death.

Miss Lipe said: "Miss Foister and I were spending the night with Miss Walker in her Biltmore home. It must have been 3 o'clock in the morning when the telephone rang. Miss Walker answered it. The call was from my sister who said the water in the river was rising fast and would soon be over the bridge."

Miss Nellie Lipe, Miss Foister and Miss Walker at once put on their clothes and went over to see if they could aid Captain Lipe.

Miss Nellie Lipe was pushed over to her home in her wheel chair. When the party reached the house the water had risen six inches over the street level and was boiling through the cement rail of the bridge.

Miss Lipe gave up her wheeled chair to her aged grandmother who otherwise could not have been moved and calmly watched her trundled to safety. Later several men successfully battled with the current and dragged a baggage truck from the Biltmore station to the Lipe's door. On this the brave lame woman was taken back to Miss Walker's home.

"After I was beyond the danger zone," continued Miss Lipe, "the remainder of the party again returned to urge Captain Lipe to abandon the house. That is all I know about the night. My sisters and father did not come back and I waited alone for daylight with the water swishing through the lower rooms.

"In the morning, I could just see my father and the two nurses clinging to the third tree from the Biltmore lodge gate. I couldn't see my sister, Kathleen, but I later learned she, too, was struggling in the tree.

"My father did not strap my sister to the tree. She tied her sweater around the trunk and attempted to work her way up the branches as the water rose. I don't remember seeing anybody actually let go, though the two Miss Walkers disappeared first. Then my father and sister went.

"Some men came for me in a boat about four in the afternoon. They made a landing on the stairs about half way up. They took me to Biltmore hospital where my sisters were brought later."

It was R. J. Dowtin and Zeb Creasman, of Biltmore, who finally rescued Mrs. Milholland from the tree in which she stayed from early morning until 3 o'clock in the afternoon. Numerous fruitless efforts had been made to save Mrs. Milholland, and she had been greatly aided by the efforts of young Mr. Thompson.

Mrs. Milholland reached a tree about 60 yards southwest of the Biltmore administration office when overwhelmed by the flood. Before the water had risen very high Mr. Thompson carried a rope from a tree directly below the office to the one Mrs. Milholland was in. The water, however, rose so rapidly that he was unable to get her ashore or get back himself. But he succeeded in tying her to a tree. Later some lumber drifted against the tree and they climbed up on it.

Mr. Dowtin and Mr. Creasman launched a raft from the office, a rope being tied to the raft, and they succeeded in getting the raft near enough to the tree so that Mr. Dowtin could jump out on the lumber which had drifted against the tree. Mr. Dowtin then succeeded in bringing Mrs. Milholland onto the raft which was then pulled back to the Biltmore office by means of the rope. Mr. Dowtin stated that the stream was so swift that he did not think they would ever have succeeded in rescuing Mrs. Milholland except by means of the rope.

George Digges, police desk sergeant, with Adolph Marquart and Harry Noland, secured a canoe and put into the Swannanoa river at the broadest expanse of the swollen stream near the lower Victoria road at the end of the Biltmore avenue car line, with the purpose of rescuing from the half-submerged Lipe home on the Biltmore side and of the family who might remain.

The large crowds which from early dawn thronged the highway and banks of the stream in an eager effort to lend help to the flood victims watched with deep interest the daring deed. From the bank well out into the flooded area of meadow land the canoe made good progress but on reaching the main strong current of the river capsized and it was not seen again. The three young men were thrown bodily against a tree and sought refuge in the limbs but again were hurled into the water, the tree giving way beneath their weight. Spectators on the northern bank of the stream witnessing this felt that all hope was lost, but each of the three being good swimmers and, as Sergeant Digges put it, "not knowing what they could do until they had to," swam with the current to the new concrete Biltmore bridge and were lost to view. From the bridge the brave trio made their way to the Lipe house and finding it utterly empty, began their perilous return. This distance was made in three stages, each more difficult than the preceding one until finally the seemingly impossible was accomplished when the boys "shinned" their way up one of the telegraph poles left standing in the river, which carries the sole remaining telephone connection to the stricken village, and on the wire strung across the river made their way, hand over hand, suspended above the boiling current.

In the last stage of this difficult progress the swirling waters beneath and the height of the wire combined with an overcoming exhaustion caused the boys to lose their hold. Falling again into the river they were at last within reach of their friends on the bank. Ropes were thrown them and amid the applause of those witnessing the daring feat they were brought to shore.

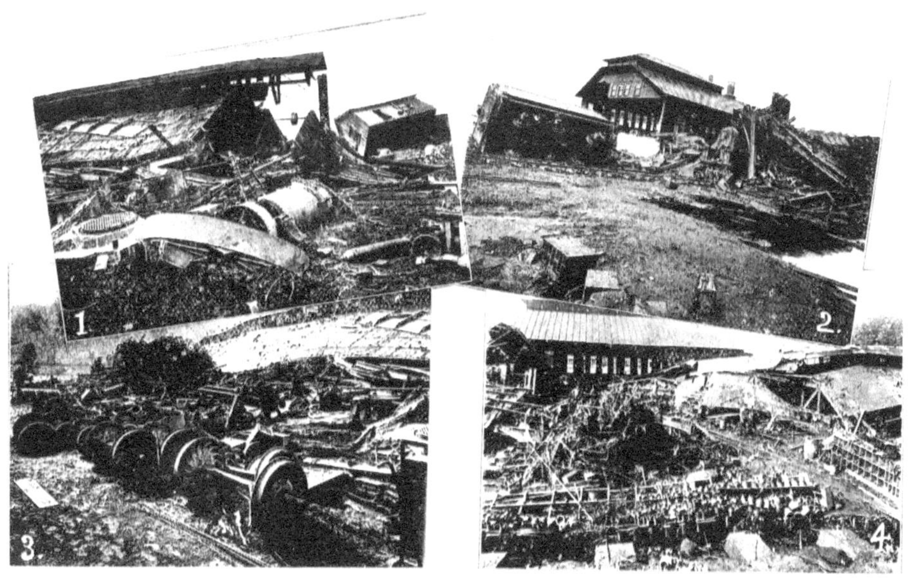

THE CAROLINA MACHINE CO., ASHEVILLE

Showing four views of this big machine shop completely wrecked. The swiftness of the river through this plant, and its powerful sweep, may be gathered from view No. 2. Here is shown a tender to locomotive swept from the railroad tracks and turned over.

NURSES MET DEATH LIKE HEROES.

Details of the drowning of Misses Charlotte and Marion Walker and Miss Mabel Foister reveal that those unfortunate victims of the flood met death with a courage as pathetic as inspiring.

Those who saw from high ground in Biltmore the tragedies that they were powerless to prevent say that the trained nurses met their fate with even more than the heroism that one would expect from a woman trained to face death from day to day in the operating room or in the stillness of the ward at midnight.

Miss Foister had been spending the day and night with her friends Miss Charlotte and Miss Marion Walker at their apartments on the Plaza. Miss Charlotte and Miss Foister both were graduates of Clarence Baker Memorial hospital at Biltmore, Miss Walker in the class of 1911 and Miss Foister in the class of 1913. All three young ladies were friends of the Lipe family, and when awakened between three and five o'clock Sunday morning by the general alarm, and hearing that their friends, the Lipes were in danger, they hurriedly dressed and hastened to their assistance.

Walking or wading ankle deep in water from the first, they had made three trips from the Lipe house to their apartment, carrying what they could carry of the household goods, bedding and linen to their rooms. When they had made the third trip, and at about 6 o'clock, the water which had been slowly rising all the time suddenly came down like a tidal wave, with terrific force and swiftness, and the young women were caught between the Lipe home and the Plaza, and carried down towards the Lodge gate. They caught hold on a large tree surrounded by one of the steel fences by which all the shade trees in and about Biltmore village are protected, and to this metal framework they clung with their hands, in water up to their necks, and with the tremendous pressure of the current pulling and sweeping against them.

Their predicament could hardly have been more terrible, more appalling to the stoutest heart, but the brave girls still kept up their courage and their hopes. Knowing that they could not hold on long, and that to let go was to be instantly carried away and under by the swift-rolling waters, and with no immediate succor in sight they still tried to cheer each other. Miss Charlotte Walker was the first to fail in strength, and Miss Foister was seen to hold up her friend's head with one hand, maintaining her own hold on life and safety with the other, till the current tore the girl's weakened grip from the metalwork around the tree and she was instantly swept out of sight. Miss Foister then supported Miss Charlotte's little sister, Miss Marion, until the little girl could hold on no longer and was torn away and lost to sight in the flood. Miss Foister was not able to maintain her hold long after her friends were swept away. Before the agonized eyes of scores of watching, helpless people, hoping against hope that she could hold out till help came, the unfortunate young woman was at length suddenly seen to let go and almost instantly sink in the muddy waters.

Horace Smith, of Beaverdam, an eye-witness of the rescue of Elmer Bishop from the flooded water of the Swannanoa Sunday afternoon, told the story of the occurrence as follows:

"Young Bishop and a young man named Fletcher had come down to the river on a motorcycle Sunday afternoon. Hundreds of people were massed on Biltmore avenue watching the rescue work on the other side of the river. Mr. Paperone had been trying to cross the river with the help of a rope thrown across a telephone pole, and had found the current too strong for him.

"Mr. Bishop and Mr. Fletcher got off their motorcycle and Mr. Bishop waded out into the stream saying, 'I can wade to the street car waiting room out there.' Dozens of people urged him not to attempt this, and to every step he took, called to him to some back, that the current was too strong for him; but he went on as if their voices were the wind, deeper and deeper into the muddy water. Suddenly he took one step too far, and was caught by the current and swept away.

"I never saw anything like the strength and swiftness of that current. That boy seemed to be turning somersaults in the water. He was rolled along head over heels, sometimes out of sight. We thought he was gone. He was impeded by his clothes, including a heavy coat and leggings.

"People kept calling to him to steer to the right, in the hope that he would get into shallow water and out of the worst of the current. He did manage to do this, and presently we saw him stop and come to his feet.

"Do you know what saved his life? He was above the Vanderbilt nursery, and he caught hold of one of those little trees, or he might have been in the French Broad now.

"Everybody called to him to hold fast, and Mr. Paperone and another man started to him. They waded all the way, but one man was up to his chin when he got there, and they were big, stout men, able to stand up in the stream. They did not have to go into the worst of the current.

"They brought him to shore, and he was the whitest live man I ever saw. He was sick from swallowing that muddy water. It was the muddiest water that ever flowed, so muddy that it was thick. It was some time before Mr. Bishop was able to get on his motorcycle and go home."

Mr. Paperone, who was born on the bay of Naples, and has swam in rivers, lakes and the oceans, stated that he had never breasted so fierce a current as that of the Swannanoa Sunday.

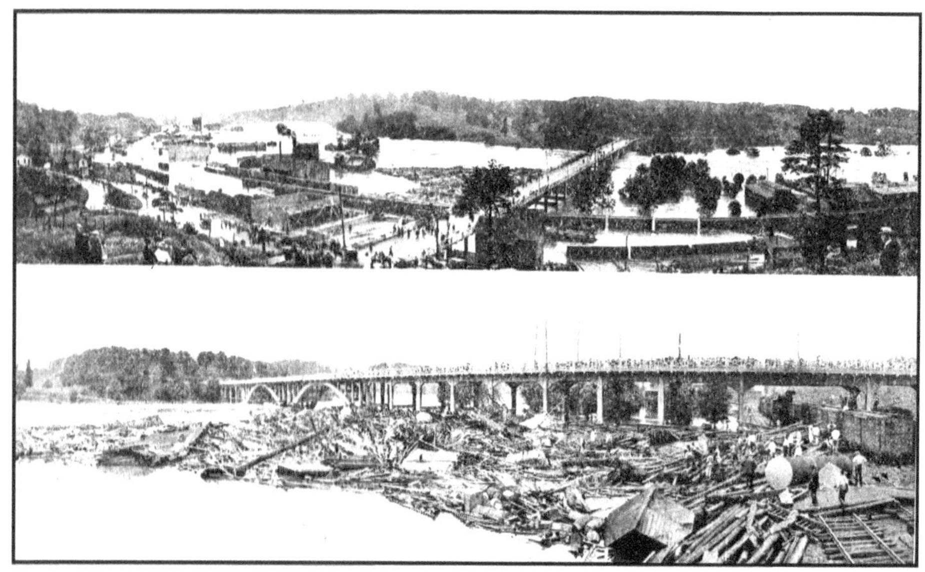

FLOOD SCENES AT ASHEVILLE.

Upper view shows bridge connecting West Asheville, and flooded railway yards and industrial section. Lower view, shows same bridge with wreckage lodged against it. This bridge stood and was not damaged.

FINDING OF THE BODIES.

The body of Miss Walker was found Monday morning, shortly after daylight, just below the lodge gate to the Vanderbilt estate.

About 1:30 o'clock Wednesday afternoon the mud bedraggled body of Miss Mabel Foister was found a half mile from the lodge gate in the Biltmore bottoms, on the Biltmore side of the river, where the young nurse had rested Sunday after her losing battle with death.

Wilburn Hendricks and Constable Joyner of West Asheville, members of the searching party, found the girl. She was placed in Mrs. George W. Vanderbilt's automobile and taken to an undertaking establishment.

The body of Captain Lipe was not found until Thursday afternoon in a muddy inlet a little farther down the Swannanoa by which he had lived so long.

MRS. VANDERBILT'S PART.

Mrs. Vanderbilt placed every resource of her vast estate—personal and otherwise—at the disposal of the relief and rescue parties.

With her daughter, Miss Cornelia Stuyesant Vanderbilt, she came early to the water's edge, and remained there, serving hot coffee to the men who were chilled by long hours in the water, offering them stimulants, encouraging the work of rescue, and putting her automobile at the disposal of those going to and fro from the hospitals in Biltmore. Mrs. Vanderbilt was dressed in a walking costume with short skirt and rubber boots, and as the men trying to rescue Captain Lipe, his daughters and the nurses, swam to shore, she would wade out and meet them, and hold hot coffee to their lips.

Mrs. Sadie Rogers, of Cincinnati, was visiting her son, Jake Rogers, near Biltmore. Mr. Rogers was called before daylight to the water's edge to try to rescue Miss Lipe. Mrs. Rogers, after waiting hours for her son to return, followed him to the scene of the flood, and was standing watching him swimming in the turbid waters, when she heard a voice at her elbow.

"Will you do something to help?"

"I turned," she said, "and looked into the face of a sweet and gentle lady."
"I'll do anything I can," I said.

"Then take this coffee percolator, and give a cup of hot coffee to every man that comes out of that river. When your coffee is gone, go get it filled from the big urn in the drug store. This lady behind you is to follow you with sandwiches."

"I said, 'all right,' " said Mrs. Rogers, "but I wondered who this lady was who seemed to have taken charge of caring for everybody. I inquired her name and was told it was Mrs. Vanderbilt.

"I served the coffee beside her all morning. When my son came out of the water, chilled and shaking, she waded out into the river over the tops of her boots to help him in, and insist upon his taking a sip of whisky or a little spirits of ammonia. When he stepped out of the water, she wrapped a warm blanket around him, made him get into her automobile, and ordered him taken to the hospital to get warm and to rest. When the boat went after Mrs. Milholland, Mrs. Vanderbilt herself placed stimulants in the boat to be given her as soon as the boat reached her.

"The next day I was going down a street in Biltmore when I heard someone call, 'good morning.' I looked around and saw Mrs. Vanderbilt. She asked me how I felt, whether I had taken cold, and whether I had slept well. I never saw a kinder lady, or one more thoughtful for others."

The "Death" Tree—Ruin in Biltmore.

The home of Captain Lipe, but a few feet from the end of the bridge across the Swannanoa, from which all the persons drowned at Biltmore fled in the fright and darkness of that early Sunday dawn, stood till Tuesday after the flood, and then collapsed and fell into the river, to be carried downstream. Had Captain Lipe, Miss Foister and the Misses Walker remained in the house Sunday they would have escaped in safety. But the house was rocking in the current; they dared not remain and thought they chose the wiser course in fleeing into the rising waters in which they met their death.

Autos and carriages plied constantly back and forth over the bridge to Biltmore in the next few days after the flood, carrying curious persons to see the ruins in the village of Biltmore, the Lipe house, and the "death tree" where Captain Lipe and the nurses had clung to the iron tree-protector, dragged at by the cold and muddy water, till they loosed their holds and were carried under the relentless waters.

Wheat, packed into the interstices of the railing of the concrete bridge, muddy curtains in front windows of the model cottages in the village, men in bathing suits, hands and feet muddy, trying to clean off floors and porches with garden hose; handsome parlor furniture and rugs standing out in the yards to be cleaned of mud and slime, the Biltmore ticket agent sorting piles of wet and muddy railroad forms, papers and tickets, the smashed canoe in which rescuers had performed the uncanny feat of paddling through the Biltmore estate office via the windows, straw and dirt left by the water in the fixtures of the street lights, these were some of the remarkable sights.

The force of the current across its entire width even in shallow water, was unbelievable. Bricks from the sidewalks of the village were strewn across the village green as if by giant hands, and the rock and gravel washed off the macadam drives was piled up here and there as if just dumped from a wagon bed.

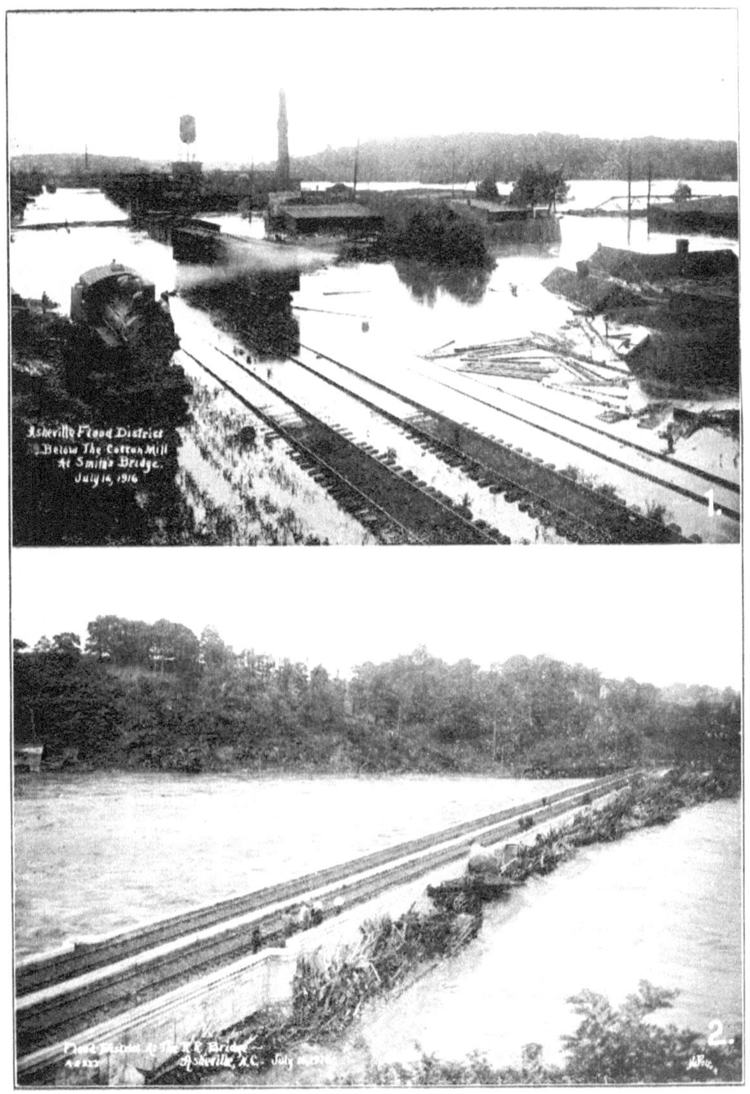

SCENES AT ASHEVILLE.

(1) Below the Asheville Cotton Mills. (2) Southern railway bridge. This bridge was left standing uninjured.

Two unknown heroes in Biltmore were a man who swam into the Biltmore livery stable and cut loose the horses so that they might save their lives, and a resident of a cottage who came back into the rising water at the risk of his life to release his six-weeks-old puppy which was calling to him from the coalhouse with piteous whines. He carried the little dog upstairs, and shut it in a bedroom with bread and water, till the falling of the river allowed him to return for his pet.

Mrs. Vanderbilt was greatly distressed at the death of the heroic young girls who met their fate in the floods, and especially at the drowning of Captain Lipe, for 27 years employee and trusted head carpenter on the estate. Following the receding of the waters she directed the work of cleaning and renovating the Biltmore cottages which showed the gruesome water mark on their walls, Mrs. Vanderbilt was also a heavy loser in respect of the hay and wheat and oat stacks from the bottom lands of the estate which were washed away, and 50 acres of fine corn buried in sand. For several days the Biltmore dairy had no ice nor water, and had to make a 10-mile detour to bring the famous "Biltmore certified" milk to Asheville.

Rescue Work at the Tannery—Death of Trexler and Frazier.

The rescuing of the marooned people in the tannery district called for many deeds of heroism.

The Asheville fire department worked untiringly from early hours until late in the evening at rescue work, and it was owing to their persistent work that every inhabitant of the Hans Rees Sons tannery district was taken out safely.

Fred Gash and Everett Frady of the fire department, Andrew Line, a boiler maker employed in the shops of the Southern railway, and Fred Jones, plain clothes policeman were the heroes of the tannery rescue party who deserve special mention for risking their own lives to save the helpless women and children marooned far out from the shore, clinging in the upper stories of rickety sheds, filled with cord wood and tan bark. Every now and then one of these buildings gave way to the pressure of the water and float out into the vortex, to be brought up with a crash and splintered against the concrete railroad trestle.

As early as 7 o'clock in the morning, when the first reports reached the center of the city that the river had reached dangerous proportions, Plain Clothes Policeman Fred Jones went to the tannery district and obtaining a boat along the river's edge made two trips alone into the dwelling section of the tannery bringing out as many people at a time as he could safely carry. Hearing that there was a sick man in the village he made for his house and found G. W. Carson, helpless in bed with typhoid fever. He rescued him and a six-months-old baby. Jones made another trip and brought out some women, but had to give up further work until more help arrived.

The firemen were summoned to the scene with the hook-and-ladder truck and one of the motor trucks.

Sladen-Fakes & Co.'s motor truck make a record run to Riverside Park, loaded up canoes and row boats, went to the flooded districts about the passenger depot where they distributed the boats. The firemen hastened to the tannery district, proceeded to rig up ferry ropes and guides to enable the boats to make safe passage across the yard and to the buildings at the far end of the group. The river was rapidly rising and the current increasing in velocity every second.

Commissioner of Public Works J. G. Stikeleather and J. H. Wood, chief of the fire department and district passenger agent of the Southern railway, worked with the men to get a guide rope from a group of loaded cars on the far siding of the railroad yards, to the nearest bark shed.

Men and women were calling for help from the windows of the second stories of the process house about a hundred yards away, but the water was rising so rapidly and flowing with such force that it was suicidal to attempt to cross with a boat. A telephone wire running from the top of the bark shed to the process house answered the purpose of a lead for the main guide rope. This wire was cut and the end of the rope fastened to it, then the marooned people in the process house hauled the rope to them and made it fast to one of the beams.

Everett Frady volunteered to take the boat across to them and pulling over-hand worked the boat out into the middle of the stream.

The water was rushing with such force, however, as to make it impossible for him to keep his position in the boat. Suddenly it was washed from under him and swirled down the river, leaving Frady clinging to the rope. He steadied himself for a moment and then worked his way over to the process house.

Another boat was immediately procured and Fred Gash prepared to make the same trip. He looped a piece of rope around the guide rope and fastened it securely to the bow of the boat, so that he could use his hands freely in manipulating the boat.

Several of the firemen and railroad men stationed themselves on top of the coal cars and handled the guide rope there, making it fast around the brake rod, while Commissioner Stikeleather and Chief Wood with others of the department stationed themselves on the first bark shed. Edward McDowell, foreman in the department, took some of his men to the second bark house, while Frady and Gash operated the boat. Thirty-five people—men, women and children—were taken out safely with this arrangement and it was thought these were all when it was discovered that three men, two women and a baby were marooned in the process house, a brick building located in the far side of the tannery grounds. This was fully a quarter of a mile from the coal car on which the guide rope had been fastened. The men now started to work their way over to these people. Gash and Frady fastened their boat to the side of the process house and crossed on the roofs of the buildings until they reached the nearest building to the brick. They cut a hole in the roof of this building, and Frady cut a guy wire leading from the metal smokestack and swung over to some drift wood and ties which he used to construct a walkway between the buildings. Frady taking the baby in his arms, and Gash and the men leading the women, they crossed the roofs of the buildings and the improvised bridge, until they reached the windows of the process house where the boat was tied. A second boat brought out the other woman, now drenched by a downpour of rain. The men were brought off in the last boat.

W. W. Velliness of Norfolk, Va., a visitor in the city, was standing on the bank at the Glen Rock hotel when a canoe capsized with Lonnie Trexler and Luther Frazier, a colored boy. Stripping off his clothes as he ran, Vellines plunged into the angry stream and swam 75 yards to where Trexler was struggling for life, but the unfortunate boy sank before his rescuer arrived and did not rise. The bodies were recovered by the firemen and unavailing efforts were made to restore life.

Trexler and Frazier had gone to carry food to the Glen Rock. Hundreds, helpless, saw the men go down in the muddy waters.

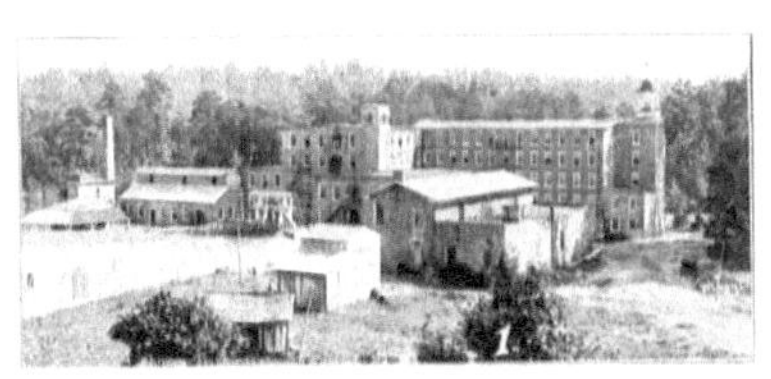

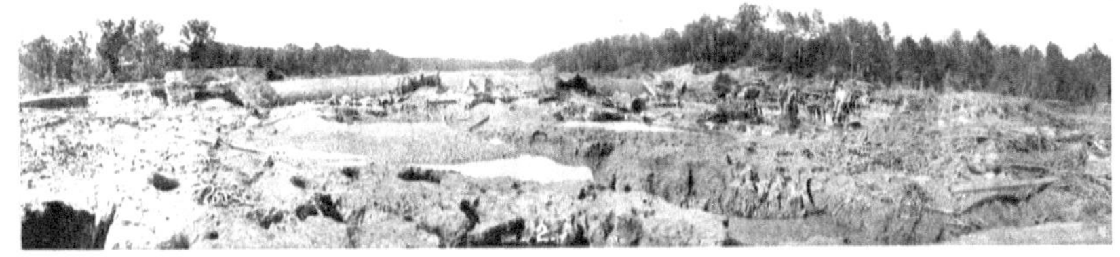

TWO VIEWS OF THE ARMON COTTON MILLS, ON THE CATAWBA,
AT MOUNT HOLLY.

(1) Upper view shows mill before the flood. (2) Lower view shows site of mill
swept clean by the flood waters, leaving only twisted machinery and piles of sand. In
addition to this loss, 700 bales of cotton went with the cotton warehouse.
(Photos by The Moons, Charlotte.)

DESTRUCTION OF RIVERSIDE PARK.

Riverside park, one of Asheville's playgrounds, is no more. Trolley tracks,
stone foundations, masses of debris, even an overturned car were jumbled
together in wild chaos. Riverside lake is a muddy pond now, separated from
the swift rush of the river by a few bedraggled trees, clinging to a mass of sand
and ruins.

The iron bridge connecting Bingham's school with Asheville disappeared.
Not even the concrete abutments remain. The several houses about the lake
are strewn over an acre of ground in a thousand pieces. The valley as far as one
can see looks like a pile of desert sand covered with flood-washed rubbish.

A strange feature of the action of the water is demonstrated along the
double line of tracks crossing just below the lake to the park grounds. The steel
rails held together in spite of the rush of the stream. They lie half buried among
a mass of wreckage stronger probably than the original dam. Not a vestige of
road ballast remains.

Reuben Newton and his family came perilously near to losing their lives in
the first rush of the flood. Mr. Newton owned a house situated at the eastern

end of the Bingham bridge. The current split about the structure cutting it entirely off from Asheville proper and threatening each moment to tear the bridge from its foundations.

At nine o'clock Sunday morning Mr. Newton, his wife, his son and only daughter and their negro cook were in the house. They failed to realize their peril until all chance of reaching the high bluff on the east bank of the river had passed. Already the bridge was straining and groaning under the power of the stream.

Mr. Newton ordered his family across the bridge, as a last desperate resort. As the party of five rushed from the structure on the opposite shore there was a terrific crash and the bridge crumbled into the torrent.

The Newtons saved a centerpiece from their dining table and one valueless photograph.

The road connecting Riverside park with the Weaverville highway resembles a gravel pile. Only the larger stones remain, every particle of earth being swept away.

By actual measurement, the flood water submerging the park reached the ten-foot mark above the trolley line.

Thomas Settle, who resides at "Orton" overlooking the French Broad river above Riverside park, was an eye-witness to the destruction of the Newton house on the river at the Asheville end of what was formerly Pearsons bridge. Mr. Settle said that he did not get to the river in time to see the bridge at Riverside park go down but was present when the Newton house floated down stream.

A remarkable feature of the rescue of the family of Mr. and Mrs. Newton was that it was accomplished across the river and that the members of the family, all of whom were saved, had crossed Pearson bridge and sought shelter on the other side of the river as they were unable to make their way from their residence to Pearson drive on account of the deep waters flooding in.

General Situation in Asheville.

Monday morning dawned on a strange and trying situation in Asheville. In the whole city and surroundings not a wheel was turning. Lamps and candles had dimly lit a night of Egyptian darkness, when for the first time in almost 30 years not a street light burned in the city. Notice was being made, not a street car, of course, was running, not a train pulled out and only one approached—a lone train on the Murphy division which came as near to the depot as the western end of the railroad bridge over the French Broad, two miles away. Railway officials could not say when it would be possible to run trains.

An alarming feature was the shortage of public supplies. Food did not at any time run low, but the kerosene and gasoline in town were soon exhausted. In fact, the police commandered the available gasoline before it was all sold out.

"There's our gasoline," said the driver of a public automobile, pointing out to sight-seers at the concrete bridge a great tank car imbedded in lumber piled up against the pier.

Newspapers had to hand-set type and get out their papers on small flat-bed presses. The telegraph offices were swamped under a deluge of messages from visitors or residents in Asheville wishing to assure out-of-town friends that they were safe, that damage and loss of life were confined to the river fronts. Western Union operators worked by 36-hour shifts, two days and a night at a time, all week, being cometimes as much as thousands of messages behind. Extra operators could not get into town and wire trouble added to the difficulties encountered.

There was no means of communication with Hendersonville, Weaverville, Black Mountain, of Marshall, except by foot.

"We have food, air and water," said the citizens. "Let's be thankful for that and go to work and clean up the flood districts and appraise the damage."

Asheville water comes from the slopes of Mt. Mitchell, 18 miles from the city and hundreds of feet above it, and unlike the situation in flooded lowland cities, the supply was never endangered.

GREAT FINANCIAL LOSSES.

Enormous losses were suffered by the industrial plants along the banks of the French Broad and Swannanoa rivers in and about Asheville. The loss of the Citizen's Lumber Company is estimated at $65,000; $50,000 in the Asheville yards and $15,000 at the Biltmore yards. The estimate of the loss at the Hans Rees tannery is said to be over $200,000. Another great loser was the Asheville Cotton Mills Company, whose loss is approximated at $75,000 to

$100,000. The National Casket Company, located on the French Broad river, is said to have lost at least $75,000. The Carolina Machinery Company's loss is estimated at over $75,000. The property of the McEwen Lumber Company was damaged to the extent of about $35,000. Williams & Fulgham lost lumber valued at about $12,000. The losses of the lumber companies at Azalea is said to be enormous. The Southern Railway Company's loss at the freight depot and in the vicinity of the concrete bridge was very heavy.

The loss at the Owl Drug store at the depot was $5,000; the stock was a total loss, the fixtures were ruined. The water stood at eight feet in the store, making it impossible to save anything.

Other plants on both rivers suffered great losses.

The property to the Southern was enormous.

It was five days before a car wheel turned in Asheville. On Thursday after the flood the Weaverville cars first came to Asheville. In about two weeks the street cars were put on half-hour schedules, in three weeks on 15 minute runs. By a strange chance, the machinery of the power houses was not ruined by the sudden flooding, but weeks of hard, patient, intelligent labor were required to clean out the sand and mud, and dry the transformers, rotary converters, and armatures.

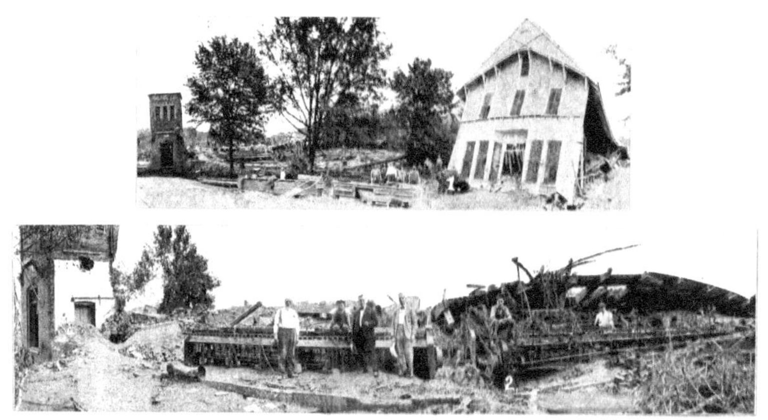

TWO VIEWS OF THE WRECKED MONBO COTTON MILLS, ON THE CATAWBA, NEAR STATESVILLE.

(1) Upper view shows part of mill tower and wrecked post office building. (2) Lower view, from another angle, showing twisted machinery and other parts of mill. Mr. Sam Steel, one of the owners of this mill is seen in this picture—the gentlemen at the left without a coat. (Photos by the Moons, Charlotte.)

A new and startling situation arose after the flood, along the water front from the railway station to the cement bridge. Thousands of gallons of gasoline, oil and kerosene soaked into the lumber piles, box cars, factories and storehouses. A carelessly thrown match, a cigarette butt, or a spark from a locomotive and a fire rivaling the flood horrors might instantly have sprung into life.

Wednesday morning the situation was most startling. A stream of gasoline a foot across was leaking from one oil tank and pools of the stuff stood everywhere. Large quantities of kerosene had also soaked into the buildings and piles of lumber.

The danger of fire was minimized by the warnings given and the strict precautions taken by the city authorities, as soon as the dangerous condition was realized.

Rescue of a Boy Marooned 48 Hours.

Tom McDowell, a 16-year-old boy of West Asheville, was rescued Tuesday morning from the store of J. C. Brice, on the west bank of the French Broad river near the concrete bridge, after having been marooned in the little building, surrounded by the flood waters of the French Broad since early Sunday morning.

McDowell had taken refuge from the rapidly rising waters in the little store, and his refuge became his prison. The store floated out into the river, where it jammed against some trees. The water rose to the boy's knees, to his waist, and Tom sought higher ground.

He stepped up on the big icebox in the store, and then bethought himself that he might become hungry before he reached dry land again, and wading and splashing about from shelf to counter he filled his pockets with boxes crackers, with fruit, cheese, and such canned goods as he could open. Thus prepared he regained the ice box and laid the eatables on the nearest shelf within his reach.

By this time the water had covered the top of the icebox, and now it began creeping up again, to his ankles, his knees, his waist.

Alarmed, the boy looked around and tried to think what he could do if the water rose to the ceiling of the store. An axe hanging on the wall opposite gave him an inspiration. Wading, half swimming, scrambling, splashing, he crossed the little room and took down the axe, and returned with it to the top of the icebox. Holding the axe above his head he chopped a hole through the shingled roof, and as the water reached his armpits he placed a box upon the ice chest, and stood up on this with his head out of the hole in the roof—and the water around his neck.

He placed his eatables out on the roof when he chopped the hole, and he coolly made a hearty meal while looking out at the scene of desolation around him. Houses, small and large, lumber, logs, a mule, washed by. Until noon, cramped, cold, faint, he maintained his critical and uncomfortable position; then the waters began to fall.

After awhile he could get down on the icebox again, then he could wade about on the floor, but still the water was too deep, the current too swift, for him to venture out. All Sunday, Sunday night, and all day Monday, he was a prisoner in his little island. Tuesday morning rescue parties out in boats heard him calling for help and took him to shore.

GRAVE SITUATION AT MARSHALL.

No single town was perhaps harder hit by the flood than Marshall, the little village on the French Broad 24 miles down the river from Asheville.

First reports from Marshall stated that practically nothing but the court house was left in this thriving little town. Fifty-three houses were said to have been carried away and two lives were reported lost.

Marshall has practically one street and its level is probably fifteen feet above normal river tide. The railway runs close to the French Broad. The main street is about 100 feet from the river.

High mountain ridges at Marshall make a narrow gorge in which Marshall lay exposed to great danger from unusually high waters.

Late reports were that the home of P. A. McElroy and the Baptist church on the upper side of Main street were the only dry buildings in the main part of town.

It is said that a telegram was received in Marshall giving warning or the coming flood in plenty of time for every one to reach safety. The three who lost their lives had been in a safe place and as in the case of inmates of the Lipe home at Biltmore, had returned to their home for some purpose, and could not again reach land.

A plank had been laid from the top of the house of James Guthrie to the cliff. Guthrie, Mrs. Estelle Bridges and the child who lost their lives, had left the house and had been safe, but for some reason had gone back. While they were on top of the house it began to move. They clung to the Chimney at the end of the house as the building moved away. The Chimney soon collapsed. Had they remained on the house they could have been saved as it washed against the southern depot and stopped.

W. A. West estimated that property losses at Marshall amounted to about $200,000. Three or four residences and some business houses were entirely washed away and parts of others were destroyed.

Mr. West stated that Shelton & Ebbs, wholesale grocers, suffered most; of their stock $8,000 to $10,000 worth floated down the French Broad river. Bales of cotton from the Capitola Mfg. Co. whirled rapidly down stream while the waters were at flood tide this firms loss being estimated at $10,000. The two banks in the town found that their records and papers were safe. The Marshall Woodworking company was a total loss.

Only two lives were lost; these were James Guthrie, and Miss Altha Briggs.

The brick building of the Marshall Motor company was undermined and one wall fell; the Presbyterian church left standing was well coated with mud on the first floors.

The Madison County Supply company, wholesale grocers, lost from $5,000 to $6,000. Others suffered heavy losses are the Madison Hardware company $4,000 to $5,000; N. B. Tweed, dry goods. $6,000; J. W. Nelson, general store, $5,000 to $6,000; R. N. Ramsey, hardware, $5,000 to $6,000; McKinney & Ramsey, general store. $5,000 to $6,000; Marshall Pharmacy, heavy loss; E. R. Tweed, general merchandise, $4,000 to $5,000; Ebbs & Halcombe, general merchandise. $5,000 to $6,000; Ebbs, Shelton & George, general merchandise, heavy loss including building.

Madison county was hard hit in the loss of a number of new bridges which had recently been completed. A steel bridge reinforced concrete bridge across the French Broad at Marshall, just finished and accepted by the county commissioner was wrecked. Three spans being carried away. A new bridge at Redmon was also destroyed. At Marshall the river cut a new channel through the large island in the stream where a great deal of money had recently been spent in laying out an amusement park.

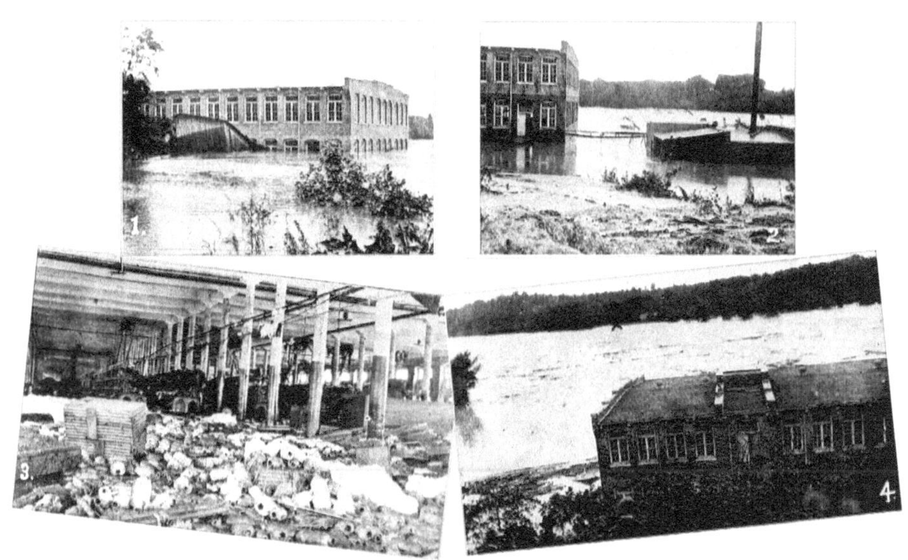

THE MONBO COTTON MILLS, ON THE CATAWBA, NEAR
STATESVILLE.

(1) Mills under water. (2) Same mill showing waters receding. (3) Interior view of mills after flood, water-marks on columns shows depth of water. (4) Mill that was completely washed away, photographed before it crumbled under the strain. (Photo Photos by the Moons, Charlotte.)

Marshall was forced to call for outside aid. Rev. Calvin B. Waller, D. D., received a letter from S. Hensley, chairman of the Marshall relief committee asking that he make an appeal to the people of Asheville in behalf of the Marshall sufferers. The appeal was turned over to the Asheville relief committee for action.

Marshall was cut off from the outside world, except by dirt roads, which were in bad condition. The letter of Mr. Hensley was sent from Marshall to Mars Hill and came to Asheville by way of Weaverville.

Mr. Hensley's pathetic letter follows;

"Dr. C. B. Waller,

"Asheville, N. C.

"Dear Sir:

"Our little town is completely wrecked, a number of people homeless, without clothes. Every store almost cleaned out, wholesale houses swept of most that was in them. We have no flower or meal. People in the country are bringing in some. No sugar in the town. We need meat and bread, cash and so forth. Will not have train for three weeks or more according to reports. If you people could help us we will appreciate it very much in this hour of great need.

"Two drowned bodies found, will be buried four p. m. Pray for us. Send any contributions to S. T. Hensley, chairman relief committee, via Mars Hill.

"Sincerely,

"S. T. HENSLEM."

CAROLINA SPECIAL, MAROONED AT MARSHALL, FIRST REACHED BY FORDS.

Five automobiles from Weaversville were the first to reach the marooned Carolina Special Monday night. The machines all Fords, were driven by Gleen West, Floyd Fox, Finley Fox, Pete Rodgers and Troy West. Two of the cars belonging to Will Reagan, two to Fred Brown and one to Marshall West. In the party was Road Supervisor Lacy of the Southern railway.

The cars made their way within two miles of the train which was stalled at Nocona, four miles below Marshall. On the way the boys overtook a large car loaded with provisions for the passengers. The larger machine out-stripped the Weaverville motors for awhile, but when it was next seen the car and the food had been deserted where the car had broken down.

Taking the food with them, the smaller cars proceeded as far as there was a road. On the return trip the machines brought four passengers each to Asheville.

The crews of the Southern railway's Carolina Special at Nacona and or two freight trains which were brought up to Nocona from further down the river reached Asheville Thursday evening coming by way of Ford automobiles through Mars Hill and Weaverville. There were 20 members of the three crews. Gleen West, Troy Fox, Floyd Fox, Peter Rogers, the same Weaverville chauffeurs that made the trip, Tuesday and brought back 20 passengers of the Carolina Special, drove the cars in which trainmen were brought to Asheville.

The chauffeurs stated that the last trip was rougher than the first and that three times on the return trip the passengers had to walk a short distance.

Mr. and Mrs. C. R. Fox of Cincinnati coming to Asheville on the Carolina Special were caught by the water Sunday at Nocona and were forced together with the rest of the passengers to take to the mountain although the train has been moved to the highest track from the water. There was fear of the dam at Marshall breaking, but by night the water seemed to be at a standstill and the passengers returned to the cars for the night, although the water was up to the platform of the cars.

Food on the dining car was very scarce and two scanty meals a day were allowed to each person. The food lasted until Monday morning. Scouting parties scoured the surrounding country and succeeded in getting enough together with the food sold to the passengers by the mountaineers to last until the next morning.

Tuesday morning the passengers walked to the dam where they found wagons furnished by the Southern railroad to take them to Walkers Gap. Automobiles were waiting there to bring them to Asheville. The trip to Asheville was the roughest and muddiest they had ever experienced Mr. Fox stated. There was only one road bridge washed away and the gully was crossed by means of planks. Only by the wonderful driving of the chauffeurs was the trip accomplished. Mrs. Fox stated that she had never seen such experienced and competent drivers as the North Carolina ones.

Mr. Fox said the only buildings left were the brick ones back from the river.

UP THE SWANNANOA. RIVER CHANGED COURSE.

The entire course of the Swannanoa river was changed at two places near Asheville, At Swannanoa and at the Cheesborough residence on the river this strange occurrence took place and as a result in the latter case a quaint old spring house, built by the great-grandfather of the present generation of the

Cheesborough family and in use for one hundred years or more, was carried down stream and utterly demolished.

Heavy pieces of furniture such as the piano in the residence of the Cheesborough family were swept down stream in the new channel formed by the flood and the lives of the Misses Cheesborough and their brother, Joe Cheesborough would not have been spared, except for the bravery of Hosea Helton and members of his family, who are of Indian descent, and have moved into the Fairview vicinity from the Cherokee reservation.

The Helton's rescued the Cheesborough family at the risk of their lives. They made a boat, and by skillful handling of the frail craft in the swift current, got the family safely across the river which had swept around behind their Cheesborough home.

F. McL. Patton made a trip on foot to his farm of 700 acres at Swannanoa to inspect the damage wrought by the storm and flood.

Mr. Patton stated that the main highway in this section was entirely washed out, the river cutting four feet or more into the roadbed and many places being made entirely impassable except on foot. Some 30 acres of land owned by Mr. Patton in the river bottom covered with a fine corn crop was destroyed. The buildings on the farm are on the mountain side and these with some 80 head of cattle remained safe. On the corn lands the river left a deposit of sand four or five feet deep and a small rivulet which runs through the farm and is a tributary to the Swannanoa river rose to remarkable proportions and cut out a gulley some 30 feet in width on its course to the river.

Several peach trees in his orchard were uprooted by landslides during the storm.

Police Court Clerk George Digges jr., left Asheville on horseback Friday after the flood following receipt of several letters from Azalea. Swannanoa and points between telling of suffering among flood refugees in that district.

Mr. Digges said that the fact that damage to farm lands in the most fertile part of the state will be great was plainly evident. The course of the flood waters along the former peaceful stream, the Swannanoa, was marked by a trail of ruined farms.

Great holes had been torn in the fields. Tons upon tons of great rocks torn from their lodging places by the water covered the productive farm lands. Massive trees, their powerful roots snapped like twigs by the flood waters, were tossed about like chips. Along the banks of the swollen stream barns and frequently dwelling houses were strewn.

Making his way through to Azalea Mr. Diggs secured a boat and crossed the turbulent waters into this former little hive of lumbermen.

The lumber plants here suffered great losses. Thousands and thousands of feet of lumber had been carried down the stream. Workingmen were engaged in cleaning out the mill and adjusting the machinery which had been tossed about the mill by the flood waters which according to residents of that village, rushed through the place with the speed of an express train.

Three fine automobiles had been caught in the flood. The machines had been anchored to trees so strange to say, were but slightly damaged. Several box cars load with lumber had been caught in the path of the water and turned over.

Several dwellings had been carried away by the flood their occupants barely having time to escape with their lives and being left helpless.

Mr. Digges, after investigating conditions, left merchandise orders for the most needy.

About a mile this side of Swannanoa the party had a thrilling experience. A large creek running out at terrific speed had overflowed its banks. The creek was still rising. Plunging in the members of the party succeeded in reaching the opposite shore of the mad stream but not without danger as their mounts frequently stepped into quicksand holes and were held up with difficulty. At Swannanoa a boat was secured and at 9 o'clock Mr. Digges started out on a search for the needy.

Conditions at Swannanoa were about the same as in Azalea. Several homes had been reduced to huge piles of broken lumber. The occupants had escaped with their lives but had been unable to save any of their personal property. The more fortunate citizens had responded nobly to the needs of the refugees and were rendering valuable aid, orders for a week's supply for the most needy were left.

Mr. Digges then started on the return trip. He was warned on nearing the swollen creek that to attempt to ford the stream at night would be tempting fate. Then started a long walk to Asheville along the roadbed of the Southern railroad.

After walking about two miles Mr. Digges secured a hand car and there began a wild ride to reach Asheville.

After two hours the parts reached Biltmore, worn out physically but happy—a day's work well done.

First News Carried By Couriers.

In the first few days after the flood, men began to come on foot into Asheville from nearby points to tell the news and to hear the news. These couriers secured copies of the Asheville papers and carried them back with them to Hendersonville, Black Mountain, Old Fort, and Marshall. They sold these papers for as much as $1.00 per copy, or, mounted on flat cars at depots or boxes in stores, read them aloud to throngs of people; or, they rented them out for ten minutes at a time, at 10 cents a "read".

W. F. Randolph, secretary of the Massonic bodies of Asheville and an old newspaper man, walked to Asheville from Black Mountain, clinging to fences and easing across shaky trestles, in his hunger for news.

Mr. and Mrs. Webb returned to Asheville Saturday, after having been marooned at Marion one week.

They went on train No. 12, which left Asheville at 2:35 p. m. Saturday. The train went as far as Nebo, and was brought back to Marion. Most of the passengers spent the first night on the train. On Sunday, the people of Marion threw open their houses, and every one had a place to stay, and had plenty of good food. A committee of citizens saw that every comfort possible was given the stranded travelers. He said the hospitality and splendid spirit shown by the people of Marion could not have been surpassed.

There was no telephone or telegraphic communications until late Wednesday afternoon.

The first train out of Marion was Thursday afternoon over the Southern Railway. Mr. and Mrs. Webb took that train to Rutherfordton; went to Spartanburg by automobile; then to Tryon by rail. From Tryon they came to Hendersonville walking and riding on work trains and then from Hendersonville by automobile.

E. B. Gresham, proprietor of Lake Kanuga summer colony settlement, made a trip from Tryon to Asheville by automobile Wednesday. The journey was attended with great difficulties, Mr. Graham stated, the road being washed away at many points. The railway on the Tryon side was in bad condition, according to what Mr. Gresham saw: at one place he said that the entire side of the mountain is gone leaving no signs of the roadbed.

Lake Kanuga did not break until about 10 o'clock Sunday morning, said Mr. Gresham. A half hour later Mr. Gresham had the news of the broken dam on the wire to warn Asheville.

W. C. Brown of the American Feed Milling company who returned to Asheville from Spartanburg Tuesday in company with five others, told a thrilling story of transportation conditions along the way.

He said: "I left Spartanburg tor Tryon Tuesday morning aboard a local train in an attempt to reach Asheville and my milling interests. The trip to the latter town was uneventful except for the evidence everywhere of water damage.

"In Tyron I discovered that all railroads and mountain roads beyond were washed away and it would be necessary to proceed either on horseback or on foot. We made up a little party of six consisting of Mr. Hanckel of Charleston, Mr. Chipley of New Orleans, H. H. Eubank of New Orleans and Mr. and Mrs. Roper of Monticello. We also took along two guides and two pack horses besides the horses we rode.

"I can hardly describe the difficulties we encountered in the mountains. In many places we were forced to dismount and lead our animals across dangerous rock outcroppings and washouts. Mrs. Roper stood the trip bravely and rivaled the men in cool nerve.

"The road was an old abandoned logging trail overgrown with bushes and blocked by boulders and landslides. At many points trees had fallen across it which made long and arduous detours necessary.

"We reached Saluda dead tired with a profound respect for the courage of the settlers of old who crossed the continent. There we left our mounts and went to Hendersonville afoot.

"To appreciate the terrible condition of the railroad right of way one must see the destruction between these two points. At one place, the river had completely washed away the embankment and the rails hung alone, suspected 30 feet in the air for a distance of 200 or more feet. All along the line we came across holes 20 to 30 feet in depth and as much as 100 feet in breadth. Telegraph poles hung at rakish angles with wires drooping, signal towers lay half buried in mud and clay and drift and cross-ties were often torn completely from under the rails and hurled down the embankment.

"I saw a large force of men working on the tracks near Saluda mountain, but their progress seemed snail-like in clearing away the masse of debrise.

"It seemed to me that the Southern will have to build a complete new bed for many, many miles—that at least two weeks must elapse before passenger service can be renewed between Asheville and Spartanburg. I want to say, however, that the Southern system is doing all in human power to take care of her passengers and put its lines in order. The company stated its willingness to me to sell me a ticket any round-about way I suggested to bring me home at the regular rate to Asheville. Considering that the railroad is the hardest hit of any of the corporations. This is to my mind, a pretty generous offer."

F. J. Ammons of Asheville, who told a remarkable story of his travels, arrived in town Wednesday from Mount Holly. He says he "skinned mules," "cooned rails," "relayed in buggies" and "hoofed it" to reach the mountain city. Mr. Ammons counted 42 washouts due to cloud burst between Rutherfordton and Asheville.

At North Thermal City the Southern tracks were washed for half a mile or more and the road-bed strewn with huge boulders. In Marion several persons lost their lives. The dead were Mr. McGee, Mrs. McGee (mother) and two children. Their house which was near the river bank was turned completely over by the flood and the family buried beneath. A prominent farmer named McKinney and his one child were also drowned. In Marion the electric lights were cut off and all water had to be brought in buckets to the city from a point a mile and a half distant. J. L. Morgan of that town lost $75,000 in cattle and crop.

C. R. Edney and R. Y. Tilson arrived in Asheville Thursday morning from Morganton, having left that place Wednesday at noon on foot for Asheville. No lives were lost at Morganton, but the Asheville men told a story of daring rescues from those who were marooned in their homes by the Catawba river. One man, Fons Duckworth was hemmed in a building surrounded by water a half mile each way to land. He was finally saved by Will Clark, who refused a reward if $12,000 for his work. Eight people were rescued from the river.

A cotton mill near the town lost 500 bales of cotton and there was some damage to industries all along the river.

Edney and Tilson told an interesting story of a family, including a little baby, carried by the flood waters or Linville river Sunday night. Next morning a searching party found the baby and its mother on a pile of debris in an eddy of the stream; both were alive and as comfortable as the unusual conditions warranted.

HEROIC WORK OF THE SOUTHERN RAIL-ROAD RESTORATION OF TRAFFIC

All this time the Southern was sending literally thousands of laborers, work cars and cars of materials out on every line in a mighty effort to restore them to normal. The work was a struggle with landslides and washouts. "The landslides did not bother us," said a road inspector. "A big gang of men with picks and shovels could soon clear them away and there was the track uninjured. But the work we did building up a new roadbed where tons and tons of earth had vanished from beneath the track and left 'grapevine trestles' hanging from cliff to cliff, was nothing less than gigantic."

Passengers were allowed to ride free on work trains, were ferried across rivers where bridges were missing, and were carried over longer routes than their tickets called for without additional charge, in the effort of the Southern Railway to do all it could for its patrons.

The damage from Old Fort to Ridgecrest was stated to be the most serious on the entire line. Here it was necessary to transport food in wagons, or light automobiles many miles over almost impassable roads to feed the gangs of men working on the steep mountain sides in isolated regions.

Over the Murphy division food, ice and mail were transported, the only outlet to Asheville and passengers journeyed in safety. Forces worked out of Asheville in the direction of Knoxville and out of Knoxville in the direction of Asheville, thus all lines in trouble were covered at the same time.

Regular train service between Asheville and Black mountain was instituted July 24.

The first train to Knoxville, after the one which raced down the river giving warning of the flood July 16, moved Aug. 5, Daily service to Spartanburg and to Toxaway were resumed Aug. 6, from Old Fort to Salisbury, Aug. 8.

MORGANTON, MARION AND VICINITY[1]

By F. C. ABBOTT

On Saturday morning, July the 15th, I had planned to take the early train via Statesville to Black Mountain to spend Sunday with my family at Robert E. Lee Hall at Blue Ridge.

The storm at Charlotte, however, had been so severe for twenty-four hours that it seemed wise to wait over another train and meanwhile consult the Southern office as to possible wash-outs on the mountain lines, and also, the weather bureau as to the continuance of the storm. Just there I "slipped a cog" for if I had taken the morning train I would have safely reached Black Mountain that afternoon and have been saved the experience which followed, and also saved my family much anxiety.

The weather man told me that the storm was moving westward and would center over Asheville that night but would clear up on Sunday. Mr. Witherspoon at the depot told me that there were no delays or trouble on the line reported. Both statements were correct.

No one now doubts that the storm centered over Asheville and vicinity that Saturday night.

At Barber's Junction about two o'clock we passed the train from Asheville which had come through that morning, which proved the statement that there was no trouble on the line up to that time.

Mr. P. S. Gilchrist joined me as we left the Charlotte station at 11:55, also headed for Black Mountain to join his family at Montreat.

Leaving Barber's Junction for the west we soon overtook the storm moving westward, and it was fully up to the standard of the rain which had just passed over Charlotte.

---

[1] NOTE—This article was published in The Charlotte News, Sunday, Aug. 6, and is reproduced here by permission.

Every little stream was a torrent, every creek a river, and when we came in sight of the Catawba it was already out of its banks and over the tops of the growing corn in the lands adjoining, and finally almost up to the level of the railroad tracks and close up to the bridge when we reached that point. We had no thought of danger, having been going up and down the road for years to the mountains. Beyond the Catawba we ran slowly over several suspicious places on the road, and finally, just beyond Connelly Springs, came to a standstill, and after a brief investigation the train was ordered back to Connelly Springs for the night.

FIRST NEWS OF TROUBLE AT CONNELLY SPRINGS.

Fifty of us picked up our grips and made a hasty rush through the train to the hotel and although it was Saturday night and past the supper hour, the electric lights disabled. Landlord Davis rose to the occasion and soon had us comfortable for the night although in some cases four in a room. It continued to rain steadily, heavily, furiously, all night long, but toward noon on Sunday the clouds began to break. We learned early in the morning that the steel bridge over the Catawba, over which we had passed a few hours before had gone down.

We had news also of the disaster at Rhodhiss mill, and that a county bridge nearby, thirty-five feet above the river had gone down and the high water mark was fifteen feet above where the bridge had been.

We tried to telephone or telegraph our people as to where we were marooned, but could only reach Hickory on the east and Morganton on the west, all wires being down from those points.

Monday, finding that we were still tied up, I called up Mr. J. A. Martin of Hickory and asked about the report as to the Rhodhiss mill, and he confirmed the news and added that several mills lower down the river were reported wrecked, and also gave us the startling news that all four bridges out of Charlotte over the Catawba were gone. We began to realize that there had been "some" flood, and to wonder when our train would proceed. A number of people from Charlotte and other points of the state were at the hotel, among them Mr. F. N. Tate of High Point who had started for Chicago and Grand Rapids. He announced early Monday morning that he was going back to High Point, by automobile forty miles to Lincolnton, then to Mount Holly, where he hoped to cross the river as a friend there had a motor boat.

His driver returned that night and reported his safe arrival at Lincolnton, and confirmed the report as to bridges gone.

Mr. Gilchrist and I discussed our situation fully and finally decided on Tuesday morning that he would make his way back to Charlotte as best he could, and so get news to both his office and mine, and I would work my way to Black Mountain, so as to advise both families that we were still alive.

I feel quite certain now that Mr. Gilchrist had the best end of the bargain, although I have not yet heard his story.

He started for Lincolnton by automobile, and I, by automobile, for Marion, where our conductor assured me that I would be able to transfer to the Asheville train, as he learned before the wires went down Saturday night that the train had come down the mountain all right. Little he knew of what had happened.

Mr. C. L. Lindsay of the Durham Traction Company of Durham, decided to join me on the trip, as his family was at Ridge Crest and would be worrying about his non-arrival on Saturday night.

We had only mild adventures between Connelly Springs and Morganton. We're stranded in a creek at one place but managed by combined efforts of driver and passengers to pull out. At another point, learning that the creek ahead was impossible we rode along the steep slope of the railroad embankment through grass and bushes. Mr. Lindsey shifted his two hundred pounds to the seat on the uphill side of the car, while I rode on the step on the same side, which helped to keep us from turning over. We found a path into an orchard, ran through this and the back yard of a farmer's home, up a lane and to the highway again, which was in very fair condition to Morganton.

Enormous Damage to Valleys Around Morganton.

Here we found Dr. Robey marooned, and learned of the enormous damage in the valley below Morganton. The beautiful valley farms seemed to be entirely ruined, buildings wrecked and washed away, land washed into ridges and covered deep with sand. Among other things stranded there from up the river was a car of furniture supposed to have been washed out from a freight train at Bridgewater eight miles away and also a large white steamer automobile with no address attached.

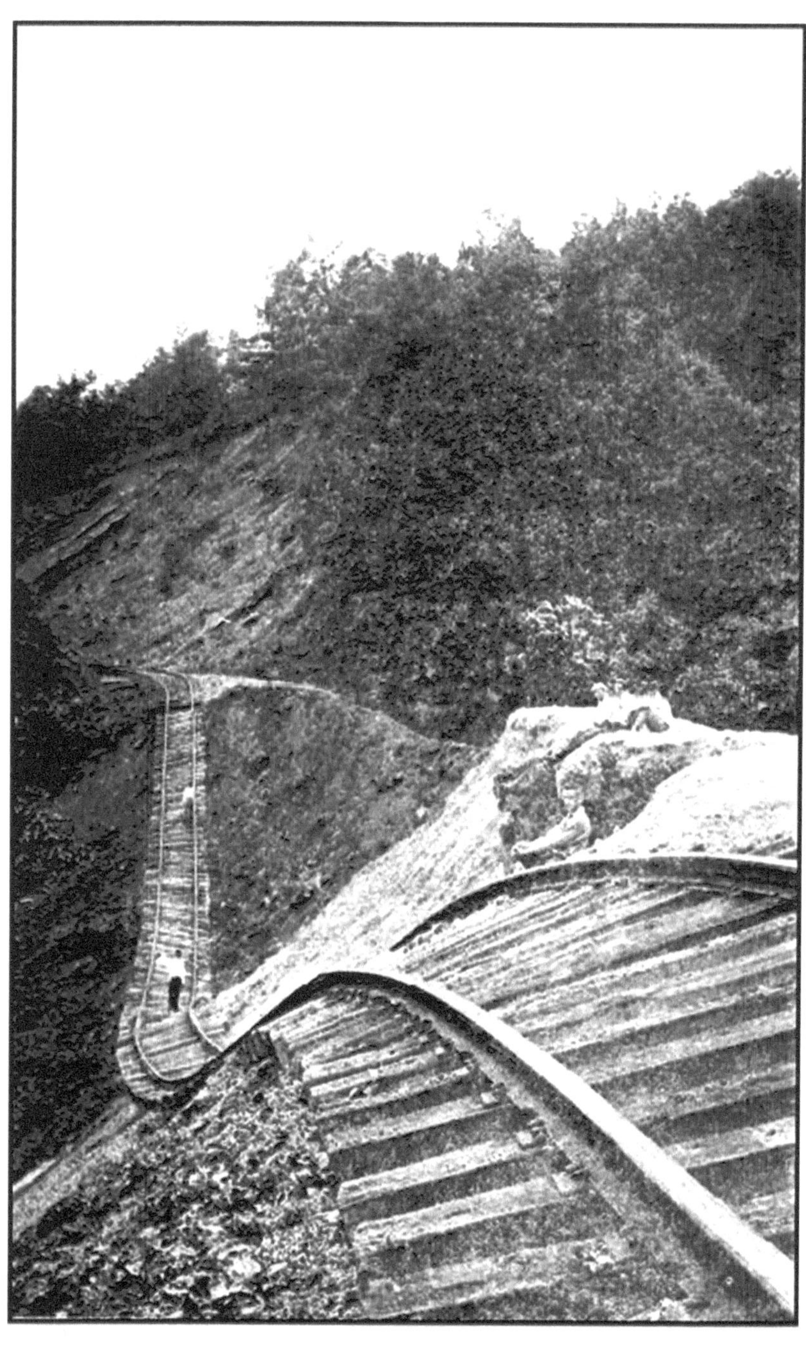

ON THE SOUTHERN RAILWAY NEAR THE TOP OF THE
MOUNTAINS.

This track hangs sixty feet in the air, looking from below.

We changed drivers and machines at this point, having been directed to the garage of Mr. J. P. Pipkin, and he at once agreed to try to get us to Marion.

Now just a line about the said Mr. Pipkin. We learned from various sources, and but little from him, that he was the hero of the occasion at Morganton. He had just come through from Charlotte in his car, during the storm, arriving home late Friday night.

On Saturday night he was awakened and told of the danger of several families down the valley. With several companions he hustled down to a planning mill and took forcible possession, over the protest of the watchman, of the lumber to build a boat. He stayed up all night and completed an eighteen foot boat, so well built that it leaked not a drop when launched into the Catawba river. All day Sunday, and Sunday night, and part of Monday, he was in the river and on the river, taking people from their flooded homes, and rescued men, women and children from dangerous positions. This was the man who had agreed to take us to Marion, and a more resourceful, energetic man of nerve would be hard to find, and his leading mechanic, Mr. Ernest Hallman, who also went with us, was a good second.

When we finally arrived at Marion we were convinced that a Ford car, with Mr. Pipkin to drive and Mr. Hallman as assistant, could be made to do about anything but climb a tree or cross the ocean.

We surmounted insurmountable difficulties, we got by immovable obstacles we drove up and down impossible places, and without a break or puncture of the new set of tires. A few illustrations I will give you of the trip between Morganton and Marion.

FROM MORGANTON TO MARION EVENTFUL TRIP.

We had no sooner crossed the bridge below Morganton, one of the very few then standing in Burke county, than we fell into a very "slough of despond." The entire road-bed over which the flood had passed was for 200 yards buried in the most slimy, sticky mud, of the consistency of thick molasses. We plunged along about 50 feet and stalled with the mud over the running gear, and thought our journey was ended almost before it had begun. But the enterprising county officials of Burke, aware of conditions, had prepared to aid travelers. A heavy mule team was close at hand and the driver came at once to our rescue. He backed up his team, threw us the end of a heavy chain, which was promptly attached to our front axle, and away we went, mule power ahead, and full power on our engine, until we ploughed our way out and took the road again.

All went well to Glen Alpine, and there we were informed that it was impossible to reach Bridgewater, as bridges were gone and two miles of the Southern track washed away, but by keeping to the hills and following old roads and fords, instead of new highways, where all bridges were gone, we might run around Bridgewater and reach Marion, where we still expected to take a train to Asheville.

"You Can't Do It, It's Impossible!" But We Did.

A countryman came along just then and said, "You can't do it, it is impossible, bridges are gone, fords impassable, you will just have to go back. Why," he added, "I tried to get across there this morning with my team and I pulled the tongue off my wagon and the neighbors had to jack my wagon out one side the creek and my team the other." Our driver took it all in, asked a question or two as to directions, and said, "Come on, let's go," and off we started. The countryman was right about the ford, and presently we fell in all right. Our wheels on one side in the quick sand almost over their tops, tilting our machine partly over on its side.

We sent for a team again, but could locate none in the neighborhood. We picked up a stout sapling near the stream, found a broad plank and a rail and with the help of a mountaineer who lived nearby, we got a leverage under the car with the sapling, raised the wheels until we could shove the plank under, scraped away a part of the sand from the other wheels with our hands, and finally by the combined efforts of the five of us, got a start, ran through the creek and up the opposite bank and were off again.

We climbed hills, where three had to push while one drove, and drove down hills on cross roads through the woods where we had to fill the ruts with old rails and limbs from trees to keep our running gear off the ground, and stranded once or twice at that.

We went over stretches of roads, which would have put the traditional rocky road to Dublin to shame. We were finally stranded once more on the banks of a creek. A mountaineer, whom we had asked for directions, had given us fair warning. "Why, you can't go through there," he said, "I tried it this morning on my mule, and he got into the mud and quick sand up to his belly, and I had to prize him out with a fence rail." "Come on, let's go," said our driver, and here we were, with the machine up to the same point in its anatomy that the mule had been, and some yards from the creek itself, which must yet be crossed.

Help was near, however, for across the stream a gang of men with a pair of oxen were already at work on the roads. One of us crossed over as best he could and asked the driver to go over to our aid, which be promptly did.

OXEN THOUGHT THE DEVIL WAS AFTER THEM.

Backing up to the machine, he hooked on and whipped up his cattle. They couldn't budge the machine, and the driver was for giving it up. "Wait a minute," said Mr. Pipkin, and sending Mr. Hallman to the rear to push, he suddenly turned on full power at low gear. The oxen jumped as if they thought the devil himself was after them, and yanked us out of the sand bar, and with ox-power ahead, gasoline power in the center and man power behind, we splashed through the creek over a high sandbar on the other side, through another mudhole and so to terra firma, again.

We drove down more hills, pushed up others by man power and gasoline power, we crossed flooded bottoms, through mud, quicksand and creeks. At one place, once a road, but now the bed of the stream, we rode up the stream over the rough stones for a hundred yards and finally, coming down a long hill to a crossing on the C. & O. R. R. we found the road absolutely blocked on the opposite side of the track.

Calling across the fields to a farmer we were told of another road about half a mile down the track which could be reached by a wide detour. Our intrepid driver said he could beat that, and sending two of our party ahead to flag any possible work train and leaving me behind as a rear guard, he turned his machine down the railroad track, and rode the ties, bobbing up and down like a rabbit, until he came to the other road half a mile away, when he called in his flagman and we all jumped in, "And the little old Ford went a running right along."

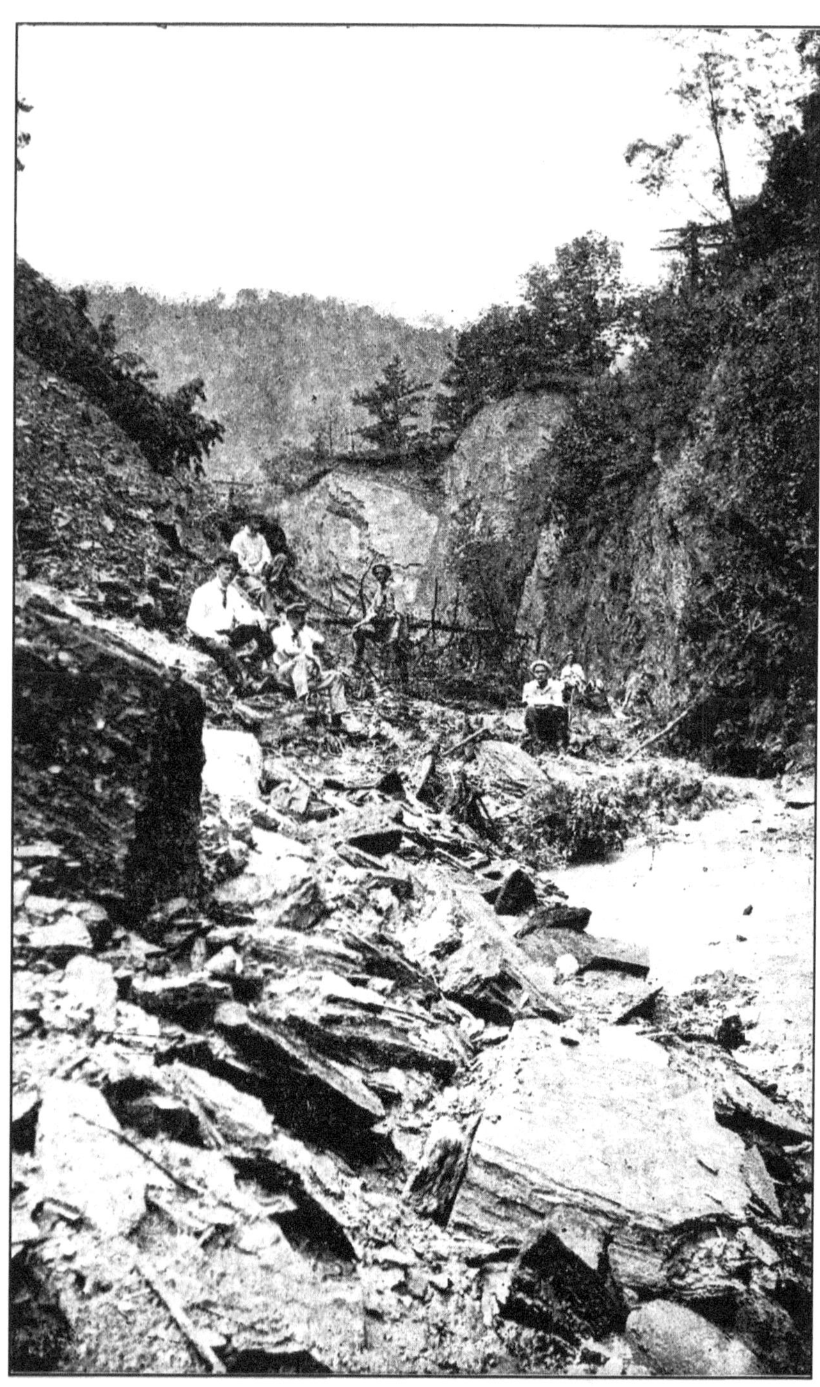

Showing Mountain Slide Completely Covering up Southern Railroad.
This Occurs Near Old Fort and Was Away from Any Stream.

We asked Mr. Pipkin what he would have done if there had been a high trestle with no ballast between the ties, and he replied he would have slipped off his tires, jacked up his car on the rails and rode the track on the rims, and knowing our man, we have no doubt he would have done it.

Soon we pulled into a little village about six miles out of Marion, and as the "shades of night were falling fast" we did not care to take chances in the dark, and stopped for the night.

REACHED MARION WEDNESDAY MORNING—LEARNS OF DESTRUCTION.

Early next morning we were off for Marion, arriving without adventure about 8 o'clock. We found over 200 people marooned there on the last train down from Asheville on the previous Saturday, and it was now Wednesday morning. The water supply of the town was off and the electric light system was also out of commission, and supplies reported getting scarce. A car of ice cream marooned on the track offered cold comfort to the inhabitants of the town.

Among the Charlotte people we met were Mr. W. C. Dowd, Mr. Ralph Van Landingham and family, Rev. J. Q. Adams and Mr. Frank Harty. Also Senator Webb, of Asheville, and Dr. W. W. Moore, of Union Theological Seminary, who had been scheduled to preach on Sunday at Charlotte.

Learning here that not only was there no train to Asheville but beyond the three mile limit, not even any wagon roads or bridges, we left our grips at the garage with cards attached and jumped into our car for the last three miles, where we much regretted to part with our automobile engineers, and started for the mountain on foot.

Mr. C. L. Lindsay, of Durham, was still with me, also Mr. Roy C. McNeill, of the Consolidated Engineering Company, of Baltimore, who wished to get to Asheville and decided to go with us.

My personal baggage from here on comprised a safety razor, a pocket comb, and a fountain pen.

The scene just above Marion was a wild one, acres of tall trees between the highway and the river bed, where the flood had passed, being bent over like a corn field after a hurricane. There were holes in the highway in which you could drop several automobiles with room to spare, a large iron county bridge at this point being quite badly wrecked. We followed the highway, or what was left of it, until we struck the railroad again and followed this to Greenlee.

Bridge Approaches All Gone—Ties and Rails Swinging.

When we came to the large steel bridge over the river approaching this place, we had our toughest experience. The bridge itself was standing all right on its piers, but 50 feet of the approach had been absolutely washed away, leaving the rails swinging in the air with the ties still hanging to them, but sagging in the center of the section about four feet, and twenty feet above the river below. I have walked many a high trestle in my younger days and it would have been no special trick to walk the ties on this swinging section of track, but the ties were in bad shape, some of them loose and hanging by a single spike and we were afraid to trust them. We learned afterward that a German citizen had crossed just before us. He was asked if the bridge was still standing, and he replied "Yes, but I possumed da rail." This was what we had to do. Those who know the graceful position of a possum on the limo of a tree this will be perfectly clear, but as some readers may not understand, I would state that to "possum" a rail, one must get down on all fours, and walk the lower flange of the rail with his toes pointed in from each side while he walks the top of the rail on his hands. Mr. Lindsey, as stated, weighs two hundred pounds, and Mr. McNeill, the Baltimore engineer, was six feet two in height and as slender as a fence rail, and the picture we presented, as we "possumed" the rail in single file, must have been more picturesque than graceful. The position was too much for me, doubled up like a jack knife, walking a slippery rail, with visions between the ties of a twenty foot drop into the river, and it got just a little bit on the nerves of all of us, and presently we changed the order of our going. We dropped on our knees, not to pray, although it seemed an opportune time, and with the rail between our knees, so that if the tier fell from under us, we would still have the rail between us and the river, we walked the fifty feet on our knees over the ties, and on our hands on top of the steel rail.

If anyone thinks this mode of travel is a joke on a hot July day, there are still plenty of opportunities over this line of railway to try it.

We reached the bridge and walked the beams to the other end, only to find that the worst was still to come. Another section of the track about 60 feet in length, and swinging twenty feet above the river, now confronted us, with a still greater sag in the center, and with also a number of the ties gone, some places two or three at a stretch. We had come too far to go back, it was now but a comparatively few miles to the end of our journey. It was case of "Pike's Peak or Bust" and again dropping on our knees, and one at a time, we took the ties as before. When we reached the vacant places where the ties had dropped off, we stretched out at full length along the rails till we could reach the next tie ahead and gracefully pulled ourselves across the gap on our stomachs, with a very striking view of the river below, until we were safely over the ties again, when we proceeded on our knees once more.

When we reached the opposite embankment, we were fully ready to quit. Mr. Lindsey with his extra weight, and entirely unaccustomed to out-of-doors exercise, was "all in." He was fairly purple in the face and his tongue was hanging out, and he sank down beside the track with the exclamation, "This captures my Angora." After a brief rest we took the track again and nothing fazed us any more, for we walked many a high trestle with the supports partly knocked out and many a swinging section of the track much higher in the air, but as the ties were still firmly spiked to the rails, we didn't even stop to consider the matter.

On the right, between Greenlee and Old Fort, a mountain range paralleled the track some distance away. There were four distinct slides visible on this mountain where rock, trees and earth had slid down the mountain from top to bottom, leaving great red gashes as if made by some gigantic plough.

HALF OF BEAUTIFUL GREENLEE VALLEY FARMS RUINED.

About half of the beautiful farming land in Greenlee valley is absolutely ruined. Arriving at Old Fort about 1 p. m. we had a good dinner and about half an hour rest. Also, by good fortune, we ran across a mountaineer, Mr. W. P. Denny, coming to Old Fort for provisions and whose home was out at Ridge Crest near the top of the mountains, and who cheerfully volunteered to guide us up the mountain. He steered us clear of many hard places, although we had all the experience we needed before we got to the top.

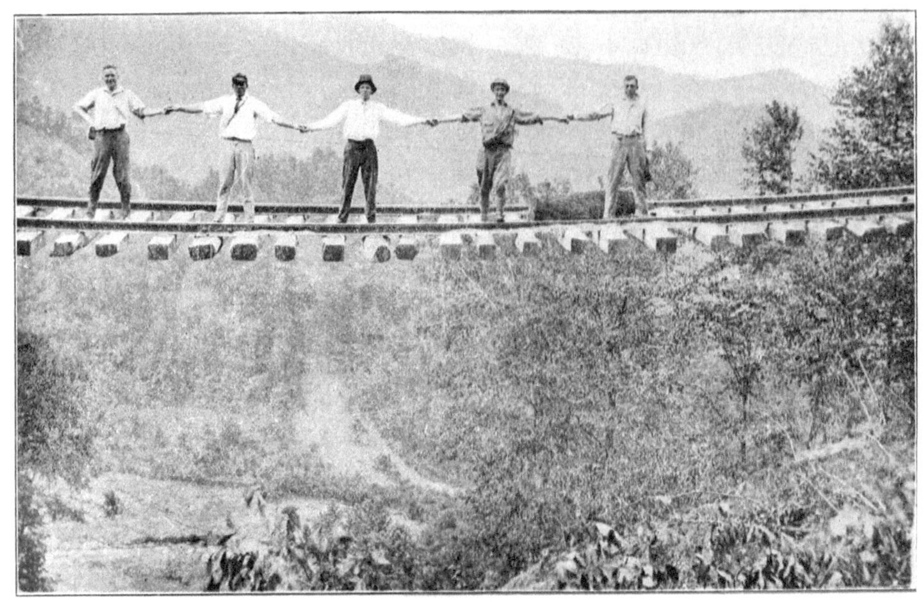

ON THE SOUTHERN RAILWAY NEAR OLD FORT.

Showing entire railroad in section away from stream of any kind. Only rails and ties left suspended high above the trees for several hundred feet.

The main part of Old Fort escaped much damage, but the new section west of the railroad was pretty badly wrecked, the river changing its course and now flowing through the center of this section of the town.

The first striking sight we noticed just as we got out of town, was a church located on the side of a hill. A landslide, and a small one at that, had come down the mountain and banged up against the rear end of the church crushing in the whole rear wall, bulged out its sides, tilted its steeple to one side, like a man with his hat over one eye, and there it stood a complete wreck.

NOT ONLY RAILROAD BUT ENTIRE FOUNDATION GONE FOR A MILE.

Around the next turn we came to the river, and a complete picture of destruction was before us. Not only the railroad but its very foundations had been swept away for the best part of a mile. Some of the track is buried under tons of sand and rock, then rises over a solid wedge of trees and stumps, then swings gracefully down in a long loop over the river to an embankment, then disappears again entirely. From here to the top of the mountain at the entrance of Swannanoa tunnel there is one continual scene of destruction. At some places track and foundations have dropped entirely into the river, heavy concrete abutments are in some cases broken and the track sagging down, several sections of track suspended in midair anywhere from twenty to sixty feet, simply the rails and ties being left, the fills having gone from under them, and in other places slides down the mountains covering the track absolutely out of sight with mud, gravel and rocks.

When we reached the first tunnel, not far from Old Fort, we found the heavy sill of a house across the mouth of the tunnel, and it was evident that the flood had not only filled a thirty foot gorge, but had flowed through the tunnel. "That's the sill of this man's house," said our guide, jerking his thumb toward another mountaineer who had joined us at Old Fort on his way home with a sack of flour on his shoulder, and later on when we had parted company with the man he told us the story as follows:

"I knew when I waked up Sunday morning after a terrible rain all night that there would be trouble down at his house, and my fourteen year old girl was down there helping his wife, as there was a three days' old baby in their home. I peered over the cliff when I came near enough and looked down on their house. One end of it was partly under water and the other end on the rocks, just a teetering up and down, ready to go down stream at any minute.

"They had all left the house, and the mother, too weak to stand, was sitting on a rock with the water partly going over her. The man with the baby in his arms was standing by, and also my own girl was there. I hurried down the mountain and got a few neighbors, went down to the convict camp, got the superintendent, who brought along a coil of rope, and we hauled them out of there. We found a pair of old wheels and rolled the woman to a place of safety."

The house, a two-story one, is still hanging there on the rocks, the little mountain farm surrounding it absolutely gone, and the stream has changed its bed to the other side of the house from where it originally flowed, so that the house can probably be occupied.

"Wait and I Will Show You Something" Said Mountaineer.

We kept exclaiming at the various sights, and each time our mountaineer guide would say, "Just wait and I will show you something." Just above the convict camp and not far above Andrews Fountain or Geyser, which has ceased to spout, we climbed out of the gorge, over a high embankment and came to a section of the new automobile highway, which we followed around the mountain for a short distance. As we rounded the last curve, our guide exclaimed, "Now look, there it is," and they lay before us a terrible scene of destruction. A section of the mountain nearly three hundred feet wide had slipped out from under the Southern railroad, leaving the track sixty feet in the air with all the ties attached, hanging in a deep graceful loop from crag to crag. Close by another slide about 100 feet in width had occurred, carrying with it the roadbed, rails, ties and all. The two slides had merged just where the new automobile highway crossed and the combined avalanche had cut through this new highway like so much cheese and had gone thundering down the valley fully three hundred feet below.

This slide has carried with it the supply pipe to the geyser at the foot of the mountain, and we saw one section of the pipe sticking out the side of the gorge. This was the most startling sight on the entire trip, and our mountaineer's description of it was most striking. He said, "They tell me a cloudburst did it, but I was just coming down the mountain and saw it done, and was very nearly caught in the slide. That water did not come from above, it came from below. The trees on the side of the mountain just popped right out by the roots, with the water spouting after them, and the whole mountain just busted wide open and slid down into the valley below."

I believe he was correct in his theory. The ground had soaked in all the water it could hold, down to bed rock, and the pressure underneath the soil on the side of the mountain had become so great that this underground reservoir simply burst out and carried everything before it.

It now began to rain heavily again, which we found decidedly refreshing after the heat of the day. About two miles up the side of the mountain, we came to the guide's home. Mr. Lindey was completely exhausted, as we had made fully 20 miles that day under conditions I have described. We helped him into a buggy standing under a shed, and he lay back on the cushions and our guide went on up to his house. Soon his two little girls came trotting down with a big can of buttermilk, and soon after our guide appeared leading his mule, saddled and bridled, and with true mountain hospitality proceeded to refresh us with the buttermilk. We then helped Mr. Lindsey on the mule and led us a mile higher up the mountain where the bottom dropped out of the road again because of another slide, and the guide and mule turned back. It now only about two miles to Ridge Crest, and on a very easy grade, and we all succeeded in pulling it through to that point, where we left Mr. Lindsey to join his family, and the two of us struck out to Black Mountain two miles away, going over more swinging sections of the track between Ridge Crest and Black Mountain.

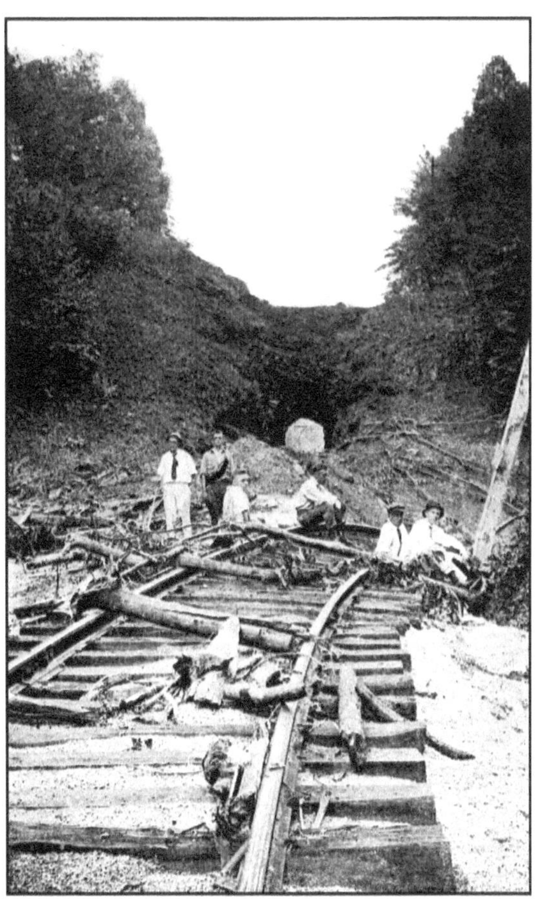

SCENE AT TOP OF THE MOUNTAIN ON SOUTHERN RAILWAY
No streams near. Deluge of water down the mountains carried away railroad ties and roadbed.

Looking Like a Tramp Reaches Destination—Slips in Back Door.

I left Mr. McNeill at Gresham Hotel and started off the remaining 2 1-2 miles to Blue Ridge in the dark. Reaching there about 9 p. m., soaked with the rain, spattered with mud, and without a collar and with shoes about gone, I sneaked into the basement door of Robert E. Lee Hall, and found my way up to Dr. Weatherford's office, and sent his assistant Mr. Jenkins, to notify my wife I had arrived. She thought he was playing a joke at first, but he finally persuaded her to come to the office, and after being introduced and finally convinced that the tramp who stood before her was the man to whom she was legally tied, she asked me how in the world I got there. I replied if I should tell her what I had done and what I had seen within the last two days, she would set me down as the greatest liar in North Carolina, but that I would tell her tomorrow after getting rested.

After resting a day and telling my story to a number of Y. M. C. A. secretaries and speakers at Blue Ridge several parties were organized to tramp down a few miles on the mountain to see some of the sights I had described. Among them Dr. Brown, of Vanderbilt University, and Dr. Kent of the University of Virginia. On their return to Blue Ridge, after witnessing the sights on this division of the Southern Railway, they told me they would be perfectly willing to sign a blank sheet and let me fill in any description I could possibly make, for it would be impossible to exaggerate the story of what they had seen. So if any of your readers see fit to question this story. I refer them to these well-known gentlemen and to any number of the Y. M. C. A. secretaries who made the trip down the mountain. In coming up the day before, our mountaineer guide had avoided the tunnels so I had not seen conditions there until this return trip. None of the five tunnels at the top of the mountain had caved in, as had been reported, for I have been through or over every one of them.

There have been land slides near the entrances of two or three of them, so that in one, some of the party waded through up to the waists in the water, and in the next tunnel the water had banked up, in one place eight feet deep, and two of the Blue Ridge men swam through it. One slide between two of the tunnels had carried away the entire railway fill to such depth that a telegraph pole was hanging by its wires adjoining the track, and the butt of the pole was fully 15 feet above the ground.

STARTS ON RETURN TRIP MONDAY—TAKES FIRST TRAIN SINCE FLOOD.

On Monday, the 24th, I started for a return to Charlotte, taking the first train since the flood out of Black Mountain to Asheville, and witnessed the destruction at Azalea, Swannanoa and Biltmore, which at the first named place is almost beyond description. The railroad to Hendersonville being still out of commission I went down by auto twenty-five miles and spent the night with Mr. A. J. Draper, at Flat Rock, who has also a very lively experience to relate about his trip from Charlotte to Flat Rock.

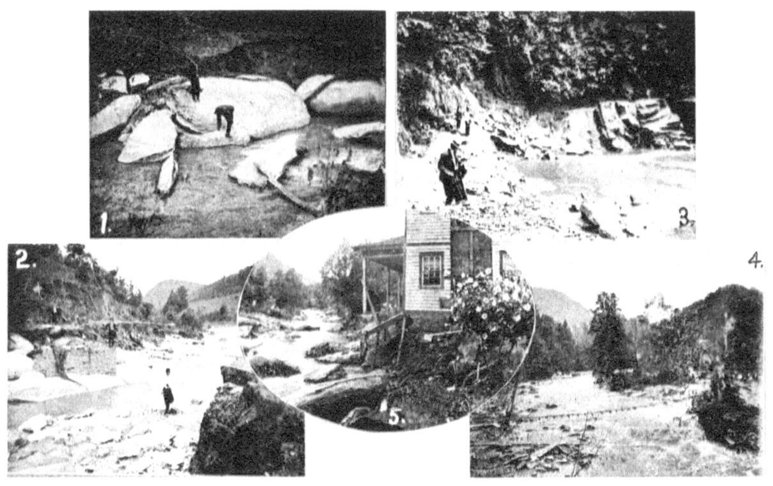

SCENES AT BAT CAVE AND CHIMNEY ROCK.

(1) Rocks washed down from mountains. (2) Abutment of a bridge, built of rock on rock foundation, washed eight feet down stream out of line of abutment on opposite bank. (3) Showing State highway built in solid rock washed away. (4) Bridge on State highway entirely gone. (5) Showing Reedy Patch creek in new channel washed away entire valley and part of home.

Early the next morning, the very first train since the flood from Hendersonville to Saluda, carried me to that point, where I stopped at Mr. S.

B. Tanner's residence to get the latest news from Charlotte. From there I tramped the ten miles down Saluda mountain to Tryon, caught a train at 5 o'clock in the afternoon for Spartanburg, and at 8 o'clock took train 40 for Gastonia and came over on the first ferry at 7 o'clock Wednesday morning, and reached Charlotte on the P. & N. the third day after leaving Black Mountain, this trip ordinarily taking six hours, and so completed the most strenuous "week-end" trip to the mountains that I shall probably ever experience.

BAT CAVE AND CHIMNEY ROCK CATASTROPHE

(Reprinted from The Charlotte News)

"Not in another hundred years, could a like disaster happen to the Bat Cave region, no matter how heavy the rains," said W. S. Fallis, chief engineer of the state highway commission, in Asheville after walking twenty-five miles through the heart of the Blue Ridge devastated by the floods of July 16.

Mr. Fallis, with Wade M. Patton, another of the state highway engineers, and just finished his inspection of the damaged highway through the Bat Cave country. This inspection was made in accordance with instructions wired him by Governor Locke Craig.

MANY GAPS.

"Out of the seventeen miles of the highway," said Mr. Fallis, "possibly five miles will have to be rebuilt. There are many gaps of from 200 to 300 feet in length absolutely gone. The bridges are all out. The terrific mountain slides were responsible for most of the damage and loss of life.

"The greater part of the damage was caused by the mountain slides. I suppose I saw the effects of more than 300 of these slides. They appeared to have started close to the top of the mountains. For a distance of possibly from seventy-five to 200 feet in which they removed everything clear and clean in their paths. It would be quite impossible to convey any idea of the terrific force of these slides. Everything movable in their path was swept to the river below. Trees were denuded absolutely of every vestige of bark. Rocks were ground smooth. Buildings were carried away in the irresistible rush. Nature had been long preparing the mountains for the catastrophe, and not for a hundred years could such another disaster happen to the mountains there, no matter how hard, or how long it might rain."

For long stretches, said Mr. Fallis, the river gorge is not more than one-eighth of a mile in width, with many sheer walls 1,200 feet and more high. During the storm from this narrow gorge an inferno of noises escaped to the starless sky above—and men who never before have known fear felt its cold hand clutch their hearts that night.

For nature once more reveled in all her ancient and elemental strength. The outcry of the river's torrent; the indescribably heart-shaking crashes of the mountain slides, one after the other; the steady and never ceasing downpour of rain, were segments of a symphony of the gods enraged—and the theme of that elemental symphony was death and destruction.

A MOUNTAIN TRAGEDY.

The highway engineer speaks of one slide, which starting slowly close to the summit of the mountain, carried away the home of E. B. Huntley. In that mountain home were the father, the mother, and their two children. Lights were burning there, for their cheer was needed, and around the hearthstone before a smouldering fire were gathered the little family. From below came the never-ceasing clamor of the infuriate driver hurling unimaginable masses of water and rocks against the mountainside. Outside a world in the making, with not a star in the heavens nor a gleam in of light. The rain came in sheets, beating against doors and windowpanes. Outside utter desolation and things they knew not of. Inside, warmth, light, fancied security.

But suddenly above the outcry of the river below was heard a still more terrific tumult above them, on the side of the mountain. It stilled all other noises, and with it came shocks which shook the dwelling and the world upon which it rested. Closer and closer came that crashing horror, and almost before the family knew of its coming it was upon them.

The man of the house staggered to the door—opened it—and in some fashion or another, stumbled outside. Before his little family could follow, and they possibly did not understand even then why they should leave that protecting glow of the smouldering fire for the utter blackness of a new world outside the slide had torn their home from where it had rested for many years, and hurled it over and over again much as a child tosses a pebble.

The husband and father, clinging desperately to a tree just outside the path of the slide, as helpless to aid or to save as a new born babe, watched with brain reeling his home with the lights still gleaming, go hurtling down the mountain towards and into that torrent of turbulent fury below, whose roar seemed to intensify in anticipation of still more victims.

The man lived—is still living—but needless to say that so long as time shall last with him never will he forget that vision of sudden death which deprived him of all that was most dear.

The mother was found later, close to the brink of the river. She was hanging, head downwards, with one foot caught in the crotch of a tree. The children were found later, too. Mother and children now rest in a common grave, close by a laurel thicket near where their home once stood.

The path of the slide is cleared of all vegetation to the living rock. Not a blade, a bush, a tree remains. In many instances so terrific was the force of that rush of earth and rocks that it possessed the characteristics of a glacier, and ground the very rocks themselves smooth. Multiply this three hundred times, in greater or less degree and the effects of these slides in the heart of the Blue Ridge mountains may be grasped, say those who have returned from that country.

In one case, at the home of J. M. Flack, the slide came down, carried off the earth upon which rested the pig pens of the owner of the farm carried pigs and pen to the bottom of the mountain and there covered them up under masses of rock and earth. But the next day the hogs had rooted themselves free of their prison and are now none the worse for their experience.

In another instance, says Mr. Fallis, the torrent excavated all the dirt from around an eighteen-foot well, leaving the well high and dry above the surrounding ground with its stone walls still intact. Instead of a well it is now a column of stone set in the midst of a boulder-strewn field.

Speaking specifically of still another instance of the flood's pranks, the engineer refers to a field completely covered with large and small boulders. This was once a fertile five-acre patch of corn. It is now covered completely with rocks, and not a vestige of dirt is visible.

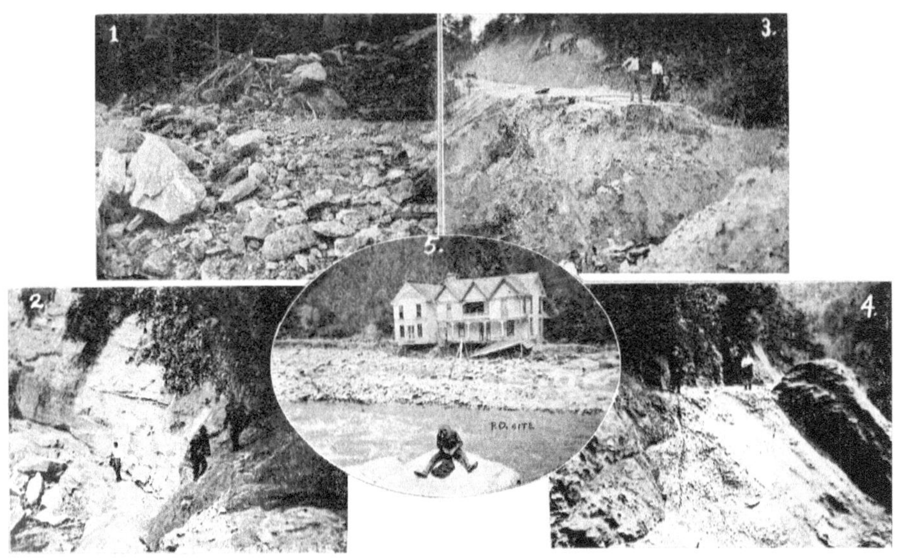

SCENES OF DESTRUCTION AT BAT CAVE AND CHIMNEY ROCK.

(1) Mountains of rock covering State highway. (2) A mountain slide of rock completely covering State highway. (3) A washout on State highway. No streams near. (4) Solid rock cut five feet deep on State highway washed away, showing hole 20 feet deep back in mountain of rock. (5) Freeman's Inn and store, only buildings left. Showing where postoffice and twelve other buildings stood. Now main channel of river.

THE TRAGIC STORY OF BAT CAVE.

Nowhere was destruction more appalling, more sudden and complete, and loss of life more horrible, than in the famous Bat Cave and Chimney Rock section;

Capt. John T. Patrick, well known as a promoter of big enterprises in North Carolina, and in recent years identified with development at Chimney Rock, arrived in Asheville from the latter point Wednesday having walked from Chimney Rock to Fairview, and coming from the latter place in a buggy. Capt. Patrick arrived attired in overhalls, and wearing but one shoe, all his clothing and other belongings having gone in the rush of the Rocky Broad river, which had played havoc with both lives and property in all that section.

Captain Patrick said the storm there began Friday, torrents of rain falling so heavily that one could see only a few yards. The destruction began Saturday, and not only was the rain heavier there than on this side the mountain, but the destruction was vastly greater in proportion to the number of homes and business enterprises involving.

"By 6 o'clock Saturday afternoon," said Captain Patrick, "the river was in full flood, and building after building was swept away, not only on the lowlands, but even on the mountain sides, where there were torrents of water filling every low place and even pouring like waterfalls down the channels 30 and 40 feet deep, from summit to valley. The landslides were numbered by scores, 25 to 200 feet in width, sweeping boulders and full-grown trees before them.

"Seven persons are known to have lost their lives in the flood, at Chimney Rock and vicinity. In one case a dwelling was torn away in which was a young woman and two children. The children were saved, but the body of the young woman, Miss Stacey Hill, was found far below the site of her dwelling, laying head down, her foot caught in the crotch of a tree.

"The horrors of that night cannot be told. The rain fell in such solid masses that one seemed to be under a waterfall, and it not only under mined houses but actually tore them to pieces. The noise of the rain was like continuous thunder, added to the roar of the river and the shock of the mountain sides literally crashing into the valleys. It was in fact a cataclysm, such as these mountains have probably not experienced in recent geographical periods. The forces of nature setting themselves to a gigantic movement simply paralyzed anything that man could do and literally stunned imagination. The people who went through that awful night can never forget the shock of it.

"Throughout the night there were hours of horror, and when daylight came the worst scene of desolation ever viewed in the mountain became visible. The river began to recede, at times, and then, strange to say, would suddenly rise again, walls of water coming down the river like an ocean tide, with the thunderous noise of waves beating on a rocky coast. The greatest height of the water was reached at between 10 o'clock and midnight Saturday night.

"At Bat Cave every store was carried off. The utter destruction of the river wiped out everything. The river has widened to two or three times its usual width. Only houses built deep in the mountain sides are standing at Bat Cave.

"The state has had for months a special force of convicts building a splendid highway between Asheville and Rutherfordton through the Hickory Nut Gap. Great stretches of this are obliterated. Bridges and high banks of earth have been replaced by holes in the ground. The aspect of the valley, in many respects one of the most scenic in North Carolina, has been in many respects changed.

Captain Patrick places the known dead at seven, but says there may be more.

Mrs. B. E. Huntley, of Bearwallow mountain, Middle Fork creek, has been found and buried," said Captain Patrick. Her children, God buried. Their bodies were not found. Miss Stacey Hill was literally knocked from her home on the side of the mountain. Two children in the house saved themselves. Issac Connor, a very old man, was at Tilton Freeman's home. They left their house to go to the barn which seemed to be on a safer site. Water undermined the barn, and as they hurried back to the house the old man got separated from them and was drowned in the flood, and a baby of Freemen, in its mother's arms was torn from her grasp, lost and never found."

Dr. L. B. Morse, who arrived at Hendersonville Tuesday night after walking with great difficulty from Chimney Rock, stated that the island at Chimney Rock was completely gone. All bridges between Hendersonville and Bat Cave and Chimney Rock were gone. Mr. Morse walked for 18 hours to reach Hendersonville and was one of the first to bring news from the Chimney Rock and Bat Cave section.

According to Dr. Morse the flood situation at these places was alarming among the buildings destroyed were the village stores.

Relief parties were organized at Hendersonville and started for Bat Cave.

Telephone connection with Bat Cave and Chimney Rock was impossible. Many telephone and telegraph wires, including those to Fairview and small villages along the Swannanoa river which were operating for a short period last night were down.

The Asheville-Charlotte highway near Bat Cave and the scenic road from the main line to the base of Chimney Rock was completely washed away. Dr. Morse was one of the owners of the scenic road which is reported to have cost $25,000.

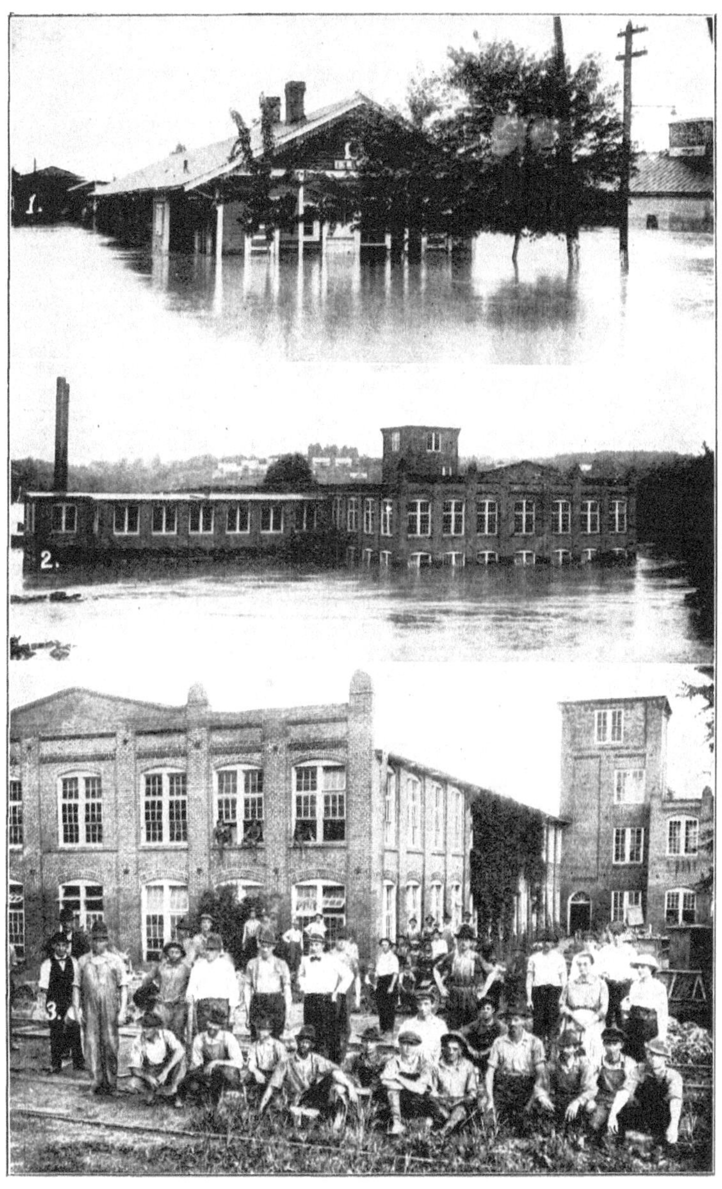

FLOOD SCENES AT ELKIN.

(1) Southern depot nine feet under water. (2) Blanket mills of Chatham Mfg. Co., half submerged. Car loaded with blankets was driven through lower end of this mill. This and water damaged mill $100,000. (3) The mill after the water went down. Force of men who have been cleaning out mud and wreckage, headed by president H. G. Chatham (standing fourth from left) and Capt. G. T. Roth (on left end, wearing black vest.)

THE YADKIN RIVER, FLOOD AND DESTRUCTION

By W. M. BELL

This Yadkin river rises in Grandfather Mountain, near Boone, in Watauga county. It is an innocent looking spring branch for several miles down the mountain side. Other branches and brooks flow into it and as it passes Patterson, it begins to be a good sized creek. Down through "Happy Valley" it gathers other water; still it is not so large that the old time foot log has to be discarded as a mode of crossing. At Elkville, in the extreme northern part of Wilkes county it gets within the river class and a few miles farther on it is fed by Reddies river above North Wilkesboro. Here it gets its first boost. A peculiarity of this stream is that all of its tributaries flow from the north side. On the south side the valley is not so wide, being cut off by the Brushy mountains and only small streams empty into it from this side.

Twelve miles below North Wilkesboro, Roaring river, a treacherous stream formed in the mountains from three forks, known as North, Middle and South Fork, empties into the Yadkin. At Elkin, eight miles farther south the Elkin creek, a good sized stream, but called a creek, contributes its waters. Between Elkin and Donnaha, a distance of about forty miles, Mitchells, Fish and Arratt, rivers empty. These rivers head in Alleghany county and flow down through Surry county for from fifteen to thirty miles. The Yadkin has a flow through Wilkes county of thirty-two miles; it is the line between Surry and Yadkin and Yadkin, Davie and Forsyth counties. It is also the dividing line between Davidson and Rowan, Stanley and Montgomery counties. At Rockingham before reaching the South Carolina line the name of the Yadkin changes to the Great Peedee.

FORMER FLOOD STAGES.

Prior to July 16, 1916, the Yadkin river had never with a single exception been known to be higher at floodtide than twenty feet. On September 23,

1898, a cloudburst in the mountains on the north bank caused a flood tide of thirty-two feet. Twenty years, before when engineers made a survey of the Richmond & Danville Railroad (now a branch of the Southern) the highest water marks were twenty feet. The railroad was located five feet higher. In those days rains would fall, and heavy rains, for possibly two weeks. The river would rise gradually and reach its crest several days after the rain had ceased and would be an equal length of time getting back to normal. The country had been developed and built, since the building of the rail road, on the same basis as the railroad grade.

RAILROADS TOUCH RIVER AT DONNAHA.

The North Wilkesboro branch of The Southern Railway from Winston-Salem strikes the Yadkin river valley at Donnaha and for more than sixty miles it traverses the valley following the river bank, in some places the river water flowing along at the foot of rock fills.

From North Wilkesboro, following the river for more than twenty miles, The Watauga & Yadkin Railroad is in course of construction, its objective point being some point in Tennessee.

On July 16, 1916, after three days of heavy rains, developing into cloud-bursts in the mountains, the Yadkin and its tributaries went on the rampage. The river rose and rose rapidly, to a height of forty-one feet at Elkin and at some places where the valleys were not so wide, a higher mark than this was reached.

Back in the mountains where the natives live in the valleys in numbers of places great slides of earth gave way and completely submerged homes, farms and vegetation down below causing untold loss of life. Down the valleys the waters swept everything before them. When the Yadkin river was reached the water carried toll amounting to millions of dollars along in its wake.

The Watauga & Yadkin Railroad, already experiencing hardships in getting on a sound financial footing was wrecked and washed away almost beyond repair.

The Southern Railway all the way from North Wilkesboro to Donnaha was damaged to an amount hard to estimate. It was forced to suspend traffic entirely for ten days and it will be months and perhaps a year before the roadbed and bridges will be in as good shape as before the flood.

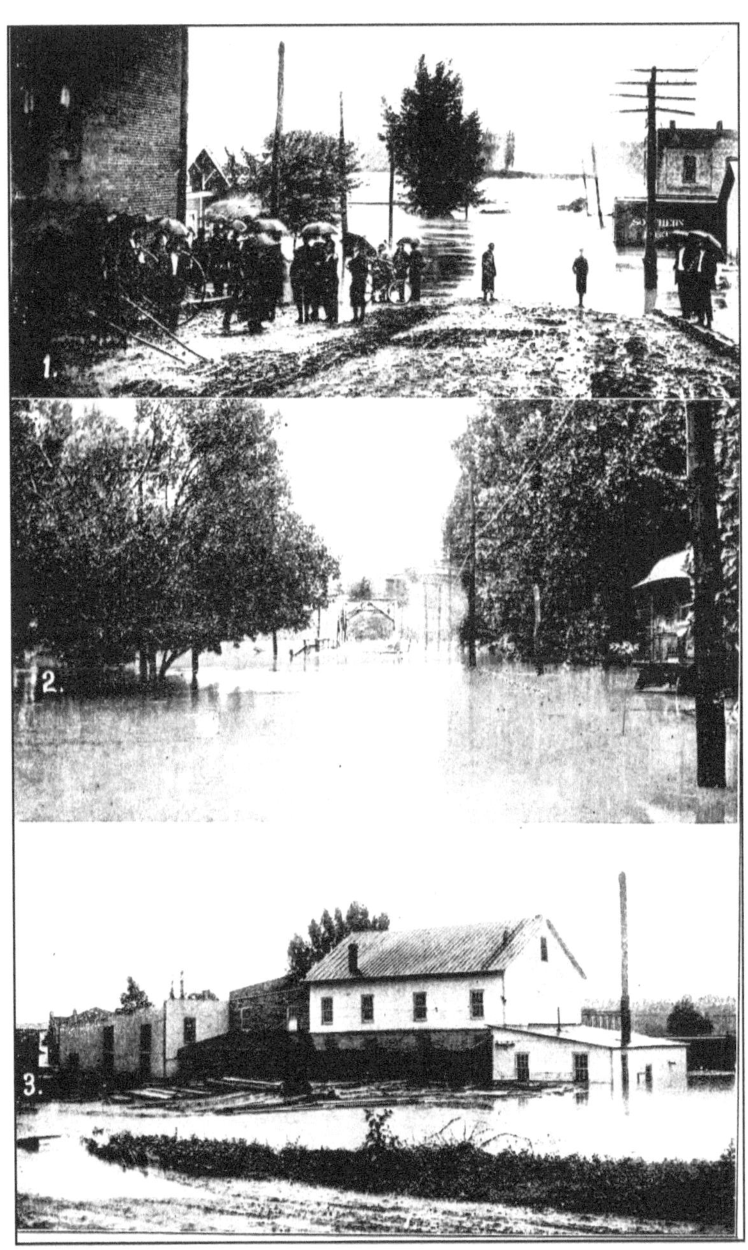

DURING THE FLOOD AT ELKIN.

(1) Bridge street looking south. On either side of this street, where the water is seen, rows of dwellings were swept away. (2) Front street, bridge floating and E. & A. railroad completely under water. (3) Roller mill, half-mile from river, and above the railroad, nine feet under water.

North Wilkesboro Hard Hit.

North Wilkesboro, the largest town on the river was, perhaps, most damaged. Located at this town is the half million dollar tanning plant of C. C. Smoot & Sons., Co. This plant was wholly submerged and in addition to property damage the company lost thousands of dollars in hides and tan bark which were washed away. The Shell Chair Co., was a heavy loser, and numbers of lumber mills and much lumber also were greatly damaged and washed away. No loss of lives was recorded at this town but there were many narrow escapes and rescues. At Ronda, and Roaring river great damage was done. At Elkin the heaviest loss was at the blanket mill of the Chatham Manufacturing Co. The water overflowed the mill, doing more than $100,000 damage. A car load of blankets on the mill siding was moved from the track and swept half way through the main mill building. Livery stables, machine shops, and everything located below the railroad at this town was either damaged or washed away entirely. Several homes were eight and ten feet in water and a number of homes occupied by colored people entirely washed away. No loss of life was reported from this town. At Burch, Crutchfield, Rockford, Siloam and Donnaha, small towns along the river great damage was done. The Southern Railway was also a heavy loser at all stations along the line to Donnaha by its freight stations being under water causing damage to freight in storage.

Damage to Farms Great.

The damage in the towns along the line hardly compares with the damage to farms and bottom lands. Many of these wide bottoms were filled with growing corn and tobacco. These crops were swept away and in many instances the whole bottoms were carried along, leaving only sand and gravel.

Long after the outside world ceases to talk about this great flood that fell with one quick swoop on these poor unfortunate people of the mountains and swept away their savings, their homes, and their farms, many of them will be still struggling hard for mere existence.

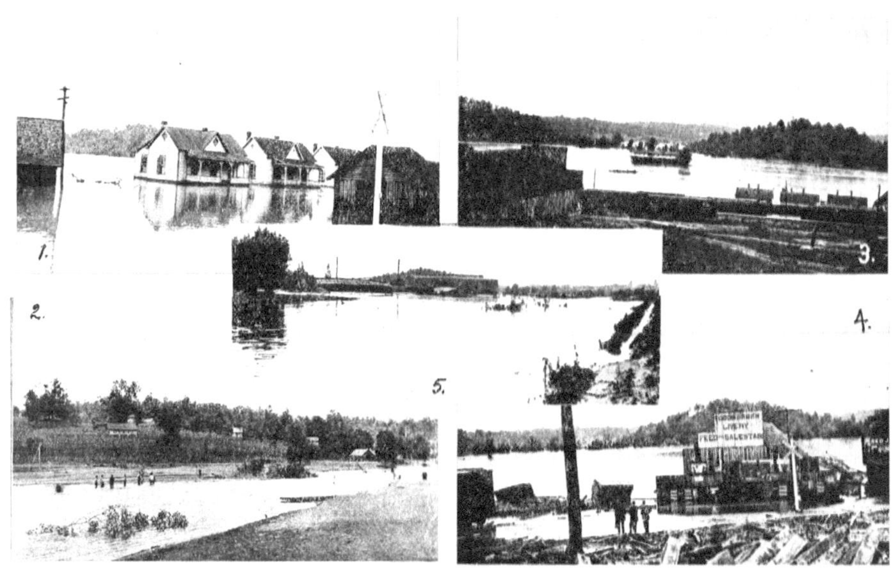

DURING THE FLOOD AT NORTH WILKESBORO.

(1) Homes under water. (2) New channel being made by Yadkin. (3) Southern railway yards showing fair grounds under water, only one building left. (4) One livery stable left in section where great lumber yards were washed away. (5) Half-million dollar tannery of C. C. Smoot & Sons., under water. "Tanner's Rest," beautiful home of Clinton Smoot, flooded to second floor.

THE FLOOD AT NORTH WILKESBORO

By ARTHUR T. ABERNATHY

As Moloch devouring the innocents; like the Ganges feeding on the bodies of breast-bereaved babes; like as Herod slaying Isreal's first-born, so did the once peaceful, placid Yadkin, lashed into a maddened fury and hounded on by the Reddies and Elk tributaries, devour in its rabid rage as it ran amuck of the fertile Yadkin Valley in the most devastating flood that has swept Wilkes county in its history.

It has been my lot to witness some harrowing catastrophes. Before the floods had subsided I stood in the Johnstown valley and witnessed the sickening spectacle of hundreds swept to their untimely graves by the bursting dam that placed that happy community in the record-book of disaster. I saw the North German Lloyd piers burn in Hoboken, and stood far out on the charred docks as the Bremen and the Saale burned to the water line, while strong men and heroic women cooped in their boiling hulks, crying to God for a succor that only came in eternity. I saw the Hunt-Wilkinson fire on Market street in Philadelphia, where sixty-six hard-working girls burned into crisps, my very blood chilled at the awful spectacle of girls making futile efforts at rescue on red-hot fire-escapes that set their clothing on fire and roasted them like festivals of demons.

But they died. They found peace.

Here on the usually peaceful Yadkin, where men work with an enterprise that is marvelous; where thrift is the household word of every family, I witnessed the little homes of families just beginning to see industrial happiness, caught in the maw of the awful storm-god, and devoured with unpitying mercilessness while brave men paced the river front where their all had gone swirling by, and cried out: "Is there no arm to save? Is there no rescue for the perishing?"

The thing that melted me to tears most as I stood on the stormy banks all night playing the searchlight of my automobile on the shaking buildings where helpless mothers and crying babies were marooned with the flood surging about them, was the absolute indifference of the men of business to their own losses in the presence of probable death to the unfortunate victims caught in the raging tide. Mr. James D. Moore, a great-souled business man whose plant had been among the first to yield its toll to the insatiate greed of the torrents, learned that two young men were about to be carried down in the Shell Chair Factory, a part of whose buildings had already been swept away. I had just retired to bed, wet and tired. Mr. Moore called me. "Can't you come with your automobile and play your powerful searchlight on the waters so we can save them?" The city lights had already been destroyed. The rain was falling in a sheet of water. I dressed in a pair of overalls—the only dry clothing I could find—and started plunging down the steep hill expecting every moment to pitch headlong down the embankment. When I plowed through the mud to the place, Moore was on the bank, building boats— utterly oblivious that he himself was one of the heaviest losers in the town. Soon Mr. C. D. Coffee, perhaps the heaviest local individual loser by the flood, drew up by my side with his automobile, and began playing its light on the Chair factory where C. H. Miller and L. E. Stacey, two employees of the Smoot tannery, had become marooned while swimming to the rescue of Mrs. Smoot and other ladies on the tanbark in midstream. The other losers joined in the futile all-night vigil—a sight not to be forgotten: brave men, facing financial bankruptcy, submerging the thought of their material losses in the dread presence of the possibility of death to the victims of the raging current.

It makes life worth the living when we realize that great industrial captains place a greater value on human life than on the mere accumulations of property.

There was a seeming fitness in the situation when the Wilkes county fairgrounds surrendered their stately structures to the gormandizing greed of the seething maelstrom. What could a fairground be to a people bereft of their fertile farms, their growing grains, their promising crops? These, too, had gone hurtling down the valley, swept away in the storm. Every iron and steel bridge in the county went with the flood, except the bridge between Wilkesboro and North Wilkesboro which remained, as if to knit us together in our common sense of desolation.

The manufacturing and lumber yard districts of this once thriving community present a sense that would sicken the stoutest heart. With mud two to ten feet deep in many places, and millions of feet of lumber entirely destroyed, the two rivers seemed to lap their tongues for more food and leaped from their accustomed courses, tearing across farms, private lawns, fertile fields and manufacturing districts until for awhile it seemed as if the entire lower part of North Wilkesboro would be engulfed.

I cannot enumerate the individual losses. Neither can I describe the details. Let it suffice to say, no pen can exaggerate them. Five persons of whom we know sacrificed their lives to the flood in this community. Others I believe will be discovered, for already vultures hover over much 0f the debris down the railroad track, which itself proved as fragile as an egg-shell before the impetuous storm. Carnegie medals should be given the two Martin boys, Oscar and Augburn, for acts of personal heroism as sturdy as ever Mucins Cordus did in the days of the Romans. They swam the stream at its height, brought in shivering men and fainting women, until they were exhausted by their efforts. North Wilkesboro will build again. No havoc can stay the progress of such a sturdy people.

IN WILKES COUNTY AFTER THE FLOOD[2]

By JOHN C. STERLING

The territory I covered embraces the village of Roaring river, the river front of North Wilkesboro, and a portion of the northwestern section of the county, known as the Reddies River section. This river and its tributaries were followed as far as possible toward their head in the Blue Ridge mountains.

The conditions I found were far beyond my expectations. The terrible havoc wrought, the untold suffering that has followed, especially among the women and little children, and the great damage done to the county of Wilkes must be seen to be comprehended in their true light.

It was my good fortune to have as my companion on the trip up Reddies River Mr. C. E. Jenkins, a hardware merchant of North Wilkesboro. He is known by and knows most of the people in his county and the roads, paths and trails are familiar to him.

Start Up Reddies River.

We got away from North Wilkesboro at 7 o'clock Thursday morning, (July 27) headed toward the Reddies river head. This river is comprised of what are known as the North Fork, the Middle Fork and the South Fork. They all come together about twelve miles from Wilkesboro, near what is known as Deep Ford. One route led up the river for a mile or more, then across the ford, and followed the Miller's Creek road to Deep Ford.

---

[2] NOTE—This story is a part of a report made by Mr. Sterling to the Relief Committee of Winston-Salem. Mr. Sterling was sent out by the committee to investigate the conditions of those in need of assistance in Wilkes county. Mr. Sterling is a member of the Winston-Sentinel staff.

Following up the river we came to the Jefferson turnpike, and up the turnpike for a short distance. Then up the river to near Whittington's store on the opposite side of the river. From this point to the head of North Fork we followed the river bed, there being no road from this point into the mountain coves and valleys.

FARMS VALUABLE BEFORE THE FLOOD.

The stream traverses a ravine between the mountain ranges. The valley almost all the way is very narrow, only little pieces of ground being available for cultivation. These little bottoms, however, before the flood, were very productive, and even away up in there were worth from $100 to $150 per acre. There were, however, very few tracts that embraced as much as an acre. The river is hardly over 50 feet in width, shallow in most places and quite swift. During the flood this river or creek extended from mountain to mountain, and so swift was the current that it turned almost the entire valley for miles and miles into a river bed leaving nothing but sand and rocks. The road up this way had followed the river's bank, and of course when the banks went so did the road. From Whittington's store up as far as we were able to get, a distance of eight miles, we were in the old road only about 150 feet. The other part of the journey was made up the bed of the river, or across fields, pastures, woods and thiskets. We crossed and recrossed the stream time and again. Some of these fords are very treacherous, not only being deep but having more or less quicksand on the bottom.

The people on this fork of the river have very small farms in cultivation. They own more or less of the surrounding mountains, but they are too steep for cultivation. The principal agricultural product of this section is corn, but they also raise wheat (about enough to furnish them bread,) quite a lot of Irish potatoes and each and every family has (or rather did have) a nice little garden which was very productive. Quite a lot of the people in the northwestern section of the country are engaged a part of the year in getting out tan bark and roots and herbs.

BIG LUMBER FLUME WRECKED.

The big lumber flume followed this stream for about 25 miles, but this flume is a total wreck, and it is said it will not be rebuilt, as the score or more saw mills have been washed away and the desirable timber pretty well exhausted. The flume was capable of transferring 250,000 feet of lumber daily from the mills up the river to the company's plant in North Wilkesboro.

There were several small grist mills on this stream, but every one of them was washed away. We found several little stores that had escaped destruction, but these stores had very little food stuff on hand.

There were a number of dwelling houses washed away or demolished on this stream. If I am not mistaken the number of dwellings destroyed numbered eighteen. Also a number of barns, cribs and other outbuildings.

Up in this territory the families are very large. Little healthy looking children are to be seen in numbers. They are very simple people in their wants and they certainly live the simple life.

RIVER BED NOW WHERE ROADS WERE.

Crossing the mountain from the North Fork we traversed the Middle Fork as far as we were able to get. This creek or river also traverses a ravine or valley. The Jefferson turnpike follows this stream some distance. The turnpike, however, is badly damaged and cannot be traversed farther than Vannoy's store owing to the fact that the river has washed the road next the mountain completely away and the river's bed is now where the road once was. About one mile farther up the turnpike leaves the river and it is said to be in fair shape across the mountain. However, the bridges back in the mountains are gone.

The damage to the river bottoms on the Middle Fork is not so complete as we found on the North Fork. However, this river brought down and deposited in fine lands great quantities of drift wood and other heavy rubbish. Trees, two and three feet in diameter, were hurled down this stream for miles. One gentleman estimates that there are 100,000 loads of driftwood on his bottom land, this gentleman being the only man in that section who has a bottom that is of any size. All this driftwood will have to be moved or burned, as it is worthless.

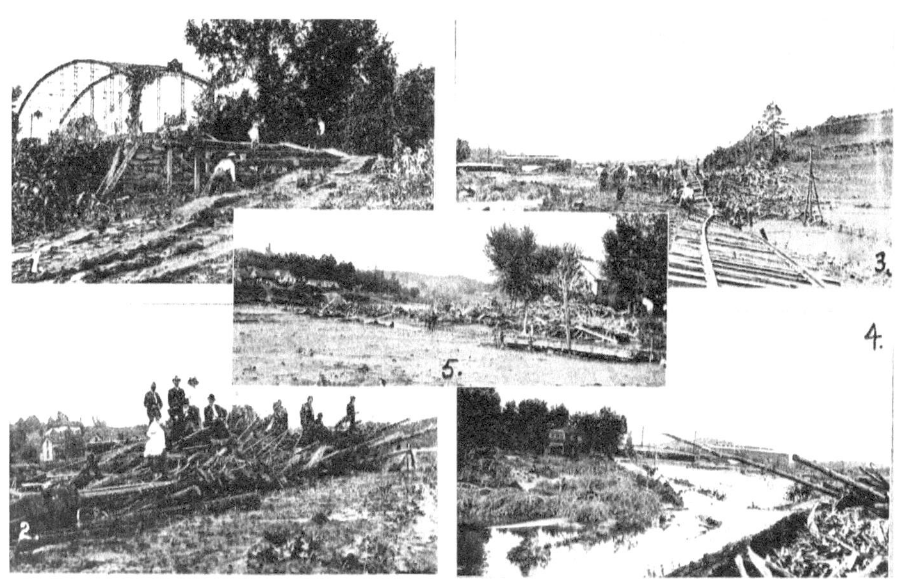

AFTER THE FLOOD AT NORTH WILKESBORO.

(1) Bridge connecting North Wilkesboro and Wilkesboro, only bridge left in Yadkin. (2) Wreckage piled on Southern railway tracks mile below station. (3) Entire roadbed of railroad washed away and piled with wreckage. (4) New Channel of river through Smoot & Sons tannery section. (5) Where homes once were.

HOMES, STORES AND MILLS ALL GONE.

We found that seven or eight houses, several stores, mills, barns and other outbuildings were washed away on this stream, and the little bottom tracts either washed away or covered over with a coat of sand varying in depth from six inches to five feet.

Following a trail across the mountain range we came into the ravine through which the South Fork wends its way. Here again we found the flood's path had ripped the valley, washed away homes, barns and every thing in its path.

HUNDREDS OF LAND-SLIDES CAUSED FROM CLOUD BURSTS.

There have been hundreds of land slides or waterspouts on the mountains in the section we visited. One party informed the writer that he could count

54 of these slides from a point near his house. These slides as a general rule started at or near the top of the mountains and carried everything before them. The width of the slide extended from 25 to 150 yards and in some instances were a mile in length. At the bottom of the mountains, where they struck, were to be found immense holes in the ground, and tons of rock and trees of all sizes. The natives report when these slides are moving they make a noise similar to thunder and come with terrific speed.

So far as we were able to learn, there were two deaths from the flood in the Reddies river section—one a little boy killed in a landslide, and the other, a woman, was drowned.

LAND SLIDE THAT DEVASTATED ENTIRE VALLEY

By W. E. FINLEY

One of the freaks of the landslides in the mountains of western North Carolina was known as the Jack Branch catastrophe in Wilkes county.

William E. Finley of Wilkesboro made a personal investigation of this landslide and wrote as follows to the Charlotte Observer:

"Yesterday I rode a horse to the top of the Brushy Mountains in Wilkes county west of Russell's Gap, tied the horse to a tree, and walked down the southern slope of Little Onion Knob to the head of a long, narrow ravine, down which flows a small stream locally known as 'The Jack Branch.' The purpose of such a journey was to see for myself that which has been the subject of conversation among all the people for miles around since the 15th of the month, the big land-slide."

WHERE DID THE WATER COME FROM?

"No one pretends to know just the source of a volume of water large enough, and with sufficient pressure to literally tear out the side of a granite cliff and hurl it with terrific force far down into the level plain below. Every one is asking, 'Whence came this ocean of water? Was it belched up out of the earth, or did it pour down from the clouds?' But no one seems to know. If the Catawba river were turned into the Jack Branch, and the Yadkin river were added for good measure, the combined strength of the two rivers would not move the huge boulders which are now lying one-half mile down the valley below where they have lain since somewhere in the prehistoric past. No one knows from whence the water came, but they all know it came, and that with such terrific force that it broke loose the solid rock from the mountain side, leaving the ragged crust of the cliff to fall in and fill up the great gap swept out by the stream of water, as if the hammer of Thor, hurled from his iron-gloved hand, had buried itself in the cliff. No sooner were these rocks broken loose than they were carried whirling down the mountain as if Neptune had pierced the clouds with his three-pronged trident and all the waters had been emptied out in the small space of 300 feet.

MOUNTAIN OF ROCKS SWEEPS DOWN MOUNTAIN.

"Beginning here, as abruptly as if blown up by a mine, a shapeless mass of debris, 20 feet high, swept down the long ravine, groaning, grinding, seething, surging to the lowlands, plowing up trees and earth and rocks as it went, and adding them to the great mass. Not only was the earth torn up to the rock beneath, but the solid rock, kept firm by the deep layer of earth covering it was chiseled out like a trough to a depth of five feet and for a distance of hundreds of yards.

"One would naturally suppose that the heavy rocks would drop out of the mass and lodge at the foot of the steep incline; but there are boulders, dozens of them, weighing at least 10 tons, lying one-half mile below where they first broke loose, which, strange as it may seem, traveled over half the distance down a grade of not more than five per cent.

"One who had never seen this valley before can only vaguely imagine the havoc wrought by this land-slide. As one looks over the desert-like waste of rocks and logs and sand, one would never dream that a week ago it had been a green valley, darkened by the shade of trees whose branches were bending under their heavy load of ripening fruit.

"Rocks, rocks, rocks! For a distance of more than a half mile along the valley, varying in width to conform to the lay of the land, there are rocks ranging from the size of coarse sand to half the size of a Pullman sleeper. They are piled and packed and jammed together in ugly confusion over all the valley to a depth of from three to ten feet. If one should venture to say that a train of 30 cars, loaded by 1,000 men, could not haul the rock from this valley and pile them up a mile away in 12 months; or if one should say that a carload of dynamite, all exploded at once, could not break loose so many rocks, he would doubtless be thought to use hyperboles. But one will be convinced that either statement would be conservative when one stands on the ground—or rocks, and sees for one's self."

"A remarkable thing about the behavior of the land-slide in its course is the fact that it did not always seek the lowest ground. For instance, there lies a mulberry tree, stripped of its bark and limbs, on the side of the hill in the path of the slide. It has been torn out by the roots and mashed to the ground. Just across the stream from it, and standing on ground 10 feet below its level is a lumber hack which was left untouched.

"There is no sign left to mark the place of the Russel home which was knocked into splinters and swept down the stream. A few pieces of furniture, or rather bits of furniture, may be seen strewn along the edge of land-slide's path. Beyond this there is no indication that there ever was a house there. A large poplar tree marks the place where Mr. Russell and his wife, each carrying a child, blinded by the mud and water, beaten almost senseless by the surging rocks and timber, fought their way to safety some hundred yards below where the house stood.

"Five centuries from now the aged mountaineers living in that region will be telling the children a story, as tradition will have it, that once upon a time the waters gathered in the mountains above and, without warning burst out in the valley and carried death and destruction in its wake. The old man will become more grave, and the eager listening child will bend his ear to hear the story of the three children who went down with the waves and of the one who was never found."

STARING DEATH FOR TWENTY-FOUR HOURS—SAVED

Experiences of KILLIAN and WHITE

Mr. J. D. Killian, resident engineer of the Southern railway who was one of the eighteen men who went down with the bridge at Maysworth, Sunday afternoon, and was rescued after twenty-four hours, by the two negroes, Alphonso Ross and Peter Stowe, speaking of his harrowing experience, a few days after wards, said:

"When the trestle was suddenly swept away I thought my time had come."

"Tons upon tons of debris and dirt of various kinds were beating against the steel structure and threatened to carry it away at any moment. The remains of a cotton mill was only a part of the debris we had to contend with.

"A steam derrick assisted us in our work. Several times the derrick moved off of the trestle to let peach trains pass. However, the last time it moved off the bridge went under.

"When the structure went down I was walking from the lower beams of one span to another. I lost my hold and dropped to the river, which was about 40 feet deep. I was sucked under and thought I would never reach the bottom. You can imagine my surprise when I suddenly bobbed up a few feet below the trestle. When I went under the terrific current caught me and shot me clear of wreckage. In the meantime the steel work of the trestle tottered over the river, seemingly just enough to let me get by safely, before it fell.

"Eighteen employees of the Southern and three linemen of the Western Union went down with the structure. Those who were not killed outright under the impact of the falling steel work were in great danger of being hit by heavy pieces of debris in midstream. We caught planks and anything else we could lay our hands upon, floated on down the river, catching hold of trees as we came to them.

"At this time the water was rising at the rate of two feet an hour and just about the time I had settled myself as well as I could in the circumstance the raging torrent would become too much for the tree and the Catawba would lift the tree by its roots and send it on its way to the Atlantic.

"In this way I was forced to change my roosting place about eight times during the night and the next morning. Once I became entangled in some vines on a tree and gave myself up for a second time when the water swept over it. Finally I disentangled myself and renewed my struggle for life.

"At dawn, H. T. Verner, and B. M. English, jr., came to the rescue of three other men and me. We were in the same tree. Verner and English succeeded in reaching us, overcoming great odds but when one of our marooned party stepped into the boat it was overturned and was swept down the river. So our little tree party was augmented by the addition of Verner and English.

"We were greatly wearied and tired out by our exertions and the two young men cheered us to renewed efforts. There were now six of us out in the middle of the Catawba. About noon on Monday, after being in the water for 18 hours, two negroes, Alphonso Ross and Peter Stowe, rowed out to us in a flat bottomed boat and took three of us to the shore. I, being in much better condition than the other three men who went down with the trestle, stayed over with English and Verner until the second trip.

"The negroes took the men safely to shore and came back for us. They told us they would take us to the shore only upon one condition, and that was that we entrust ourselves entirely to them and let them engineer the boat as they saw fit. Needless to say we agreed to the proposition with alacrity. After we were taken to the shore the negroes rescued another man who was in a tree about two and a half miles from the trestle."

Mr. Killian told of an amusing occurrence out in the river. One of the men who were in the trees was extremely nervous and on the slightest provocation wanted to float down to another tree. He was in the act of jumping off the tree he was in and making for a tree below him when suddenly one of his companions called his attention to a big water moccasin hugging the tree that was his objective. He decided to remain where he was.

Julius White, one of the colored men saved from the river Monday after 18 hours in the water, told his story in vivid language, describing the sinking of the Maysworth trestle and his long night-ride down the roaring river on rafts and logs and trees.

White said that he was working near Supervisor Griffin Sunday afternoon, cutting out the debris from the trestle, and that suddenly the trestle began to sink down. It then rose again and about that time there was a snapping noise at one end, "like all the world was coming to an end," said White, "and then I went under, holding on to a part of the track to which I had scrambled trying to get off the bridge. I went under the water, it seemed to me, at least 40 feet, and then came up."

White said that as he came out of the water he saw Supervisor Griffin and that the latter slowly shook his head at him. What the supervisor meant White did not know, but both were making towards a raft, and White says that he held back to give Mr. Griffin a chance, but that the latter suddenly went under the corner of the raft, seemingly drawn under by the current and was seen no more. Then White seized the raft and held on for his life, with two logs about his body, hampering his progress. He kicked clear of these and then went with the raft a short distance, and finally got a tree. For six or eight hours until nearly midnight he clung to trees within sight of the trestle, and finally White says "I shouted to the watchman that I was gone, and let go of the tree I was holding. I drifted down the river all night and must have been in forty trees before daylight came. It was the same everywhere. I would get a tree, and it would serve me a little while and then I would have to get out and drift on to another. I got my shoes and part of my clothes off and was thus able to swim pretty well, but several times when I tried to swim to the bank I failed because the current was too swift.

"After daylight there were two little red-birds that came to the tree where I was perched," said White, "and I looked while they seemed to be studying me. When I laid eyes on them I said, 'Little birds, I'm going to get out of here,' and sure enough it was not long till the white men came from the Gaston side in a boat and got me off that tree. It was about 70 feet from the ground where I was perched that Monday morning."

White stated that he saw the body of a woman clad in white floating down the river Monday morning. He was unable to get a very close view of the body but was certain it was that of a woman apparently drowned for some time. He also stated that he saw a cotton ginning plant in the river and could even see the new machinery in the building as it floated past. He wanted to get inside where he stated that it looked dry and comfortable but he could not let go the tree that he was holding to for his life at that moment.

THE BREAKING OF LAKE TOXAWAY

(Reprinted from Asheville Times)

A dozen times on the fateful "flood Sunday" the rumors went round that Toxaway dam had broken. It did not break that day, nor that week, nor the next, but Sunday night, August 13, four weeks after the great freshet, the dam at beautiful Lake Toxaway gave way to the long season of rain and high water, and the third and greatest lake was lost to western North Carolina.

R. F. Williams, for years operator and railway agent at Lake Toxaway wired the Asheville Times:

The beautiful lake known as Lake Toxaway is no more. On or about seven o'clock Sunday evening a small opening appeared in the dam and the water trickled through in a small stream, which rapidly grew and in a few minutes the dam was doomed.

Messages were sent to Asheville to warn towns in South-Carolina to look out for high-water, as it was feared that great damage might be done by the rushing waters. In about fifteen minutes more the whole dam fell in with a mighty crash. The water, at first clear as crystal, changed to a muddy torrent, as it rolled down the narrow chasm that nature had left between hills, carrying with it debris of all kind.

The electric power plant erected several years ago at a cost of ten thousand dollars which supplied the hotel and cottages with light went down as if it had been made of paper. But the man in charge of the plant had been warned and escaped by climbing the side of the mountain.

In the meantime the telegraph office was a busy place, and people began pouring in to wire their friends that they were safe. Much anxiety was expressed as the people in South Carolina who might be in the path of the flood.

Crowds of people watched the water as it poured through the dam. The long pent up water roared with explosions as of dynamite as it escaped behind the mighty dam that had imprisoned it so long.

The long continued rains during the summer had weakened the dam and once an opening appeared it was doomed. The hotel was not damaged in any way except by loss of the power plant supplying electric light, but the owner of the estate E. H. Jennings of Pittsburg, Pa, had suffered serious loss in the destruction of the dam which took months to build and which was constructed about 13 years ago.

The cottage people were not damaged in any way. Trains are running regularly and number of sight-seers came up to see what remained of the once beautiful lake.

There was no loss of life the property loss can be replaced and it is planned to restore the resort to its former beauty.

Just as the dam was breaking, George Armstrong of Savanah, Ga, came down the lake in his launch, but seeing the water rush through the narrow opening in the dam wisely turned around and made up stream, towards the opposite side near the hotel and reached the hotel in safety. He had to abandon his boat at the dock and when the water subsided it left the boat in a great depression in the lake. The lake is thoroughly drained and almost dry except for puddles here and there.